CARTOGRAPHER'S TOOLKIT

Colors, Typography, Patterns

CARTOGRAPHER'S TOOLKIT

Colors, Typography, Patterns

Gretchen N. Peterson

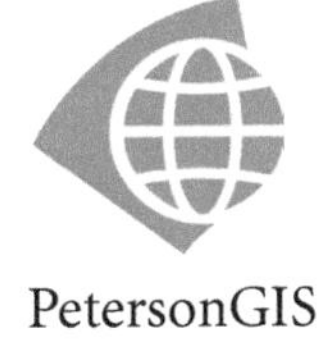

PetersonGIS

PetersonGIS

Fort Collins, CO

Find us on the web at www.gretchenpeterson.com and www.petersongis.com

Book and cover Design by Longfeather Book Design

ISBN-13: 978-0-615-46794-8
ISBN-10: 0-615-46794-6

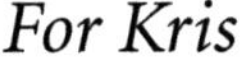

For Kris

Contents

About the Author

Gretchen N. Peterson is the owner of the geospatial analysis firm PetersonGIS, which creates custom solutions for clients in the natural resources field and produces cartography products. Peterson is also the author of *GIS Cartography: A Guide to Effective Map Design,* CRC Press, April 2009. Peterson writes a cartography blog at www.gretchenpeterson.com/blog, is on the application review committee for the GIS Certification Institute, is a co-founder of Ignite Spatial Northern Colorado, and publishes technical articles in leading geo media outlets and on www.petersongis.com. Peterson lives in Fort Collins, Colorado.

Acknowledgements

The author wishes to thank Nat Case, Learon Dalby, Don Meltz, and Amanda Taub for their editing of the Colors chapter.

For their helpful suggestions and comments on the Typography chapter, thanks go to: Richard Brocksmith, François Goulet, Hans van der Maarel, and Doug Noltemeier.

Andy Woodruff, Jim Humelsine, Gretchen Culp, Dale Loberger, and Will Stahl-Timmins all provided expert suggestions on individual sections of the book.

Special thanks to Robert and Erik Jacobson of Longfeather Book Design for book layout, Becky Dobbins of Becky Dobbins Design for copyediting and Pamela A. Anderson for copyediting.

Finally, an enormous thank you goes to the cartographers who contributed their outstanding work to this book. Your maps provide more than just examples, they are the inspirational basis that many future cartographic masterpieces will build upon.

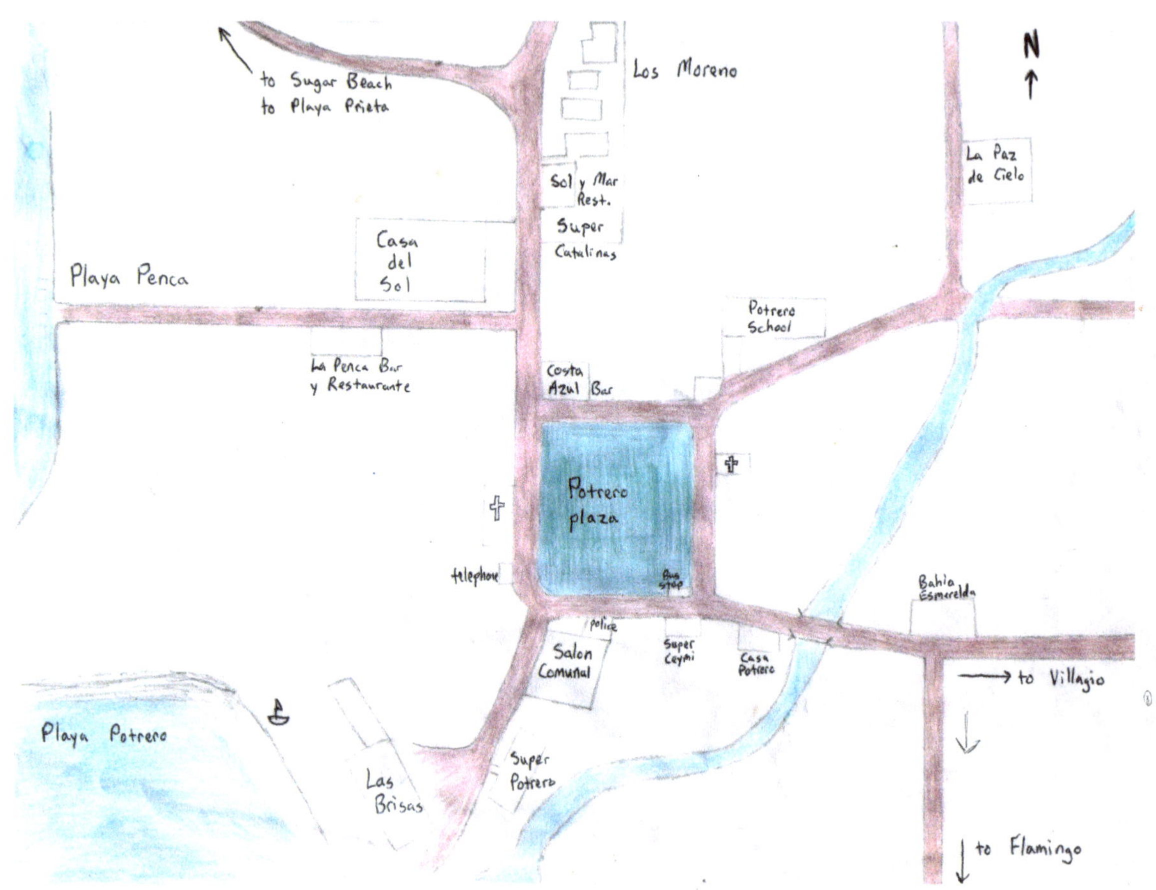

Hand-drawn map of Playa Potrero. Colored pencil tracing on Google Map from the author's personal collection. Cartographer unknown.

Introduction

In an ideal world, every cartographer would possess, at the ready, a toolkit full of map parts to assist in the map production process. Visualize this toolkit as a box stuffed with color swatches arranged in harmonic palettes, clippings of typefaces that make readable labels at tiny sizes, and a compendium of map examples from a variety of genres. When new cartography projects are thrown your way, or when you have just finished a big geo-analysis that needs to be mapped, you would rummage through the box to find the perfect parts to make your map a stand-out. This metaphorical box would comprise **the toolkit** that every map maker should have and that this book aims to supply.

For all map makers in the world—from from those who make just a few maps a year to those who design them on a weekly basis—this book seeks to: (1) inspire during times of ebbing enthusiasm, (2) emphasize that color, type, and composition are crucial to any map design, (3) broaden and deepen even the most seasoned professional's repertoire, and (4) assist the inevitable trial-and-error process through a multitude of cartography-specific examples. To meet these goals, the book contains 300 color swatches grouped into 30 palettes, showcases 25 serif and 25 sans serif typefaces, discusses 28 map composition patterns, and includes over 30 map examples by 25 contemporary cartographers.

Cartographer's Toolkit is assembled into three chapters: Colors, Typography, and Composition Patterns, building from individual map components to cohesive cartographic constructions. Each chapter begins with a brief introduction explaining relevant theory, key definitions, and usage suggestions. The pages that follow each introduction provide an abundance of visual demonstrations that are the basis for the tools in the toolkit.

The first chapter, Colors, presents 30 color palettes in each of 3 categories: coordinated palettes, color ramps, and differentiated palettes. The colors are shown as individual swatches, placed together in a color-wheel illustration, shown in different combinations (such as areas, lines and polygons), and placed in a map along with a corresponding color-deficiency simulation map. Used as a reference, this chapter will help you make maps faster and better.

The second chapter, Typography, presents 50 typefaces in each of 3 categories: standard, free, and for-fee. The typefaces are presented—in serif/

sans serif pairs—via a sample map, a description, up to four sentences, a font hierarchy, and a chart of letters and numbers. This chapter aims to broaden your typeface knowledge base as well as help you make font choices that optimize readability while enhancing the character of your maps.

The third chapter, Composition Patterns, is an information-dense exploration of map styles that includes specific introductory suggestions on uses, styles, and techniques for each. Some of the patterns describe map techniques and some describe map types. They all have one or more excellent examples to illustrate them, gathered from a diverse group of cartographer contributors, representing many parts of the world and many styles. This chapter is meant to be studied and referred to periodically. It includes enough information to help the cartographer explore each pattern further through other resources.*

There are currently a few online resources for choosing colors and typefaces for maps; however, they showcase a limited number of options. The large number of choices organized and easily accessible in this text is perfect for the serious map maker. It works for those seeking palette and typeface shortcuts too. The Composition Patterns chapter, to the best of the author's knowledge, explores map styles in a way that has never been done before. Perusing the map composition patterns, including a thorough study of their purposes and limitations, will give the reader a basis from which to launch similarly innovative and effective maps.

This book contains more tools and less theory than that which the existing body of cartography literature provides. That makes this the go-to text for choosing the elements that will make up your map. If you want to explore the myriad ways that professionals with today's technology design maps, this book provides that too. Do note that this is an advanced text that assumes a working knowledge of cartographic theory, projections, and basic geo-analysis. If you lack this knowledge, there are other cartography books that delve into map design theory that can supplement what you learn here. The author's previous book *GIS Cartography: A Guide to Effective Map Design*, is one. For projections and geo-analysis tutorials, there are many good resources online and in print.

Other supplemental resources include the companion e-books for the Colors and Typefaces chapters, which contain the same color palettes and typefaces (though the layouts are somewhat different). If creating your own book of shop-colors and shop-fonts is a goal, these e-books will provide a great starting point. They can be purchased from the author's website at www.gretchenpeterson.com/booklet.php. The Colors e-book comes with an ArcMap style file to make it easier to use these colors in that software.

Tools work best when you know how to use them. Anyone can pick out colors, use a default font, and put together a map. This book enables the thinking cartographer to use colors and fonts deliberately, and it shares ideas for creating just the right layout compositions with just the right elements to create truly communicative and enduring maps.

*In this book, website addresses (URLs) for resources are not listed, as they often change. It is assumed that the reader can search for these websites using the keywords or company names listed.

COLOR

Color

The aim of the colors chapter is to provide the map maker with a range of color palettes that can be rapidly incorporated into a map. On the outside of each color page is a series of 10 colored boxes that, together, comprise the palette showcased on that page. By flipping through the Colors chapter pages and looking just at these colored boxes, the reader can easily select a palette of interest. With ten colors in each palette it may not be necessary to use all of the colors shown on a particular page, leaving the map maker to pick and choose those that will complement the map most effectively. Palettes can also be mixed and matched, though this will take a higher degree of effort and experience on the map maker's part. One thing is certain: even the best cartographers incorporate plenty of time for trial and error where color is concerned.

Color choices in map design, as in graphic design, are often the result of inspiration. The author has gained inspiration for palettes from many sources including major works of art, web icon palettes, magazine montages, and book illustrations that were then adapted to create suitable map palettes. If the reader so wishes, new color palettes can be created using these same sources of inspiration as well as others such as photographs, graphic design pieces or even fabrics.

The advanced map maker may even put together a private collection of colors from these inspirational sources that have successfully been used in map products, to create a book of shop colors. Indeed, it is advisable to record those colors which have worked well together in the past and will therefore likely work well together in future projects. Furthermore, a book of shop colors is also advisable for teams of cartographers who need to ensure that their map products have a cohesive visual aesthetic, though strict adherence to the shop book can also cause problems if or when unique constraints present themselves. Therefore, a modified approach whereby a shop book is kept and encouraged, but not rigorously policed, is probably the best option.

The palettes in this chapter represent a range of color moods in that some are dark, some are cheery, some look like a teenager's bedroom, some like an art nouveau interior. However, the palettes are intentionally left unlabeled as it is not the author's wish to skew the reader's color preferences with descriptive words. While a color palette may have been inspired by a certain mood, the map maker may arrange the colors in such a way that the original mood is no longer apparent.

Color Reproduction

The colors in this chapter were rendered in CMYK and printed in CMYK. Keep in mind that these colors may look different on your monitor, print-outs, and production prints.

Printers

The most precise method of print color reproduction is the spot color method, most commonly specified in Pantone formulas. More widely used, however, is four-color process (Cyan, Magenta, Yellow, and Black or CMYK) used in both toner-based and ink-based machines. To expand the range of colors available in four-color processes, some systems add additional colors. For example, hexachrome printing presses and inkjet printers use orange and green in addition to CMYK. The four-color process is less exacting than the spot color method due to the fact that each color is made up of certain quantities of the other colors. With all methods, a visual inspection of the results, commonly called a press check, is helpful to ensure that the final product meets your requirements. Other factors that can affect the printed colors are the type of paper and the specific qualities of the press and ink used. Professional cartographers must be cognizant of color printing optimization procedures while other geoprofessionals may or may not need this level of color specificity.

Monitors

Ensuring color accuracy on a computer monitor is likewise difficult and inexact. It requires manual calibration or the use of specialized software and is not a one-time process. In fact, it is sometimes recommended that a monitor be recalibrated monthly. For cartographers who require an extreme level of color accuracy, these procedures are necessary. For others, the general color will be all that is required.

Color Conventions

While color theory is a worthwhile educational pursuit for the novice cartographer, the beginner with-

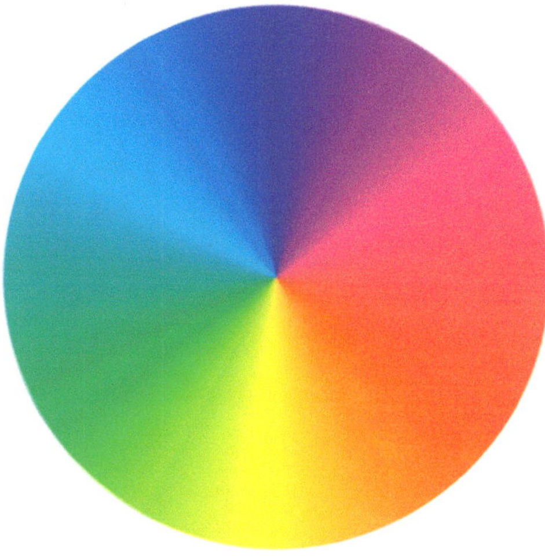

out such a foundation can have some success by adhering to certain color conventions as a shortcut method for getting started. Color conventions exist for everything from general feature types all the way to specific colors for soil and geology types. The general feature type conventions are: green, light brown or tan, for land features; blue, white or gray for water features; red, orange, black or brown for road features; and purple, pink, yellow and orange for smaller features that need to pop out of the background. Color conventions are not always followed, however. If an exception is to be made to a common color convention, then that exception should be made with full-knowledge of the way in which it differs and the reasons for it. It may also be necessary to explain this departure from color convention to the recipient of the map, be it a boss, client, or the map reader.

Examine the color wheel (left). For labeling of features there are two common methods: one is the use of a complementary color—on the opposite side of the color wheel—for the text color so that it has sufficient contrast to stand out and be extremely legible; the other is the use of an analogous color—nearby on the color wheel—or a lighter or darker shade of the same color to ensure the proper association of the feature and its label. While most color palettes in this booklet do not include pure white or black, these colors can be used on all maps, and are often used—along with gray—for map labeling.

This chapter is intended to provide general color combinations and formulas that work well together. The author does not assume responsibility for color inaccuracies.

How to Use This Chapter

THE COLOR PALETTES are arranged vertically on the outside of each page in order to allow the reader to flip through and visually scan the options. In small gray numerals next to the color swatches are number codes from one to ten, which are then used in the color combinations section —the rectangular bars of color underneath each main map—to allow the reader to match the colors with their formulas. On the other side of the swatches are the HEX, RGB, and CMYK formulas for those colors. While there are other color formula options, these are widely used formulas in GIS software and the web. Programs that can convert between the various color systems are available online.

The large map on each page uses many, but not all of the colors in the palette and is an example of how the palette might appear in a map. The map may also contain pure black or white for text and boundary lines in addition to the colors in the palette. A few of the color ramp palettes have maps that contain an extra color for the ocean area. In these cases that color is noted beneath the map.

A color-deficiency simulated image on each page provides an illustration of how the larger map would look to someone with deuteranopia, the most common inherited type of color vision confusion. Deuteranopia is a low ability to discriminate differences in the red, orange, yellow, and green portion of the color spectrum and occurs in approximately five percent of males. Vischeck software was used for these simulated images.

The circle of colors on each page is a circular visualization of the linear color palette swatches. This alternate way of viewing the colors is another tool that the reader can use to help chose a color scheme. It is only very loosely based on a real color wheel and should not be mistaken as one.

The large colored bars underneath each main map provide yet another way of illustrating how the colors will appear when combined in ways that are common to maps—contour lines, circles, roads, and labels. The numbers below each feature correspond to the numbers next to each color swatch on the outside of the page to aid in identifying which color is depicted. In a very few cases pure black (b) or pure white (w) is shown here when no other color from the palette is readable.

COORDINATED PALETTES

These palettes were selected for their harmony of color and are particularly useful for creating a coordinated map color scheme. Remember to continue their use throughout the layout including the margin elements such as legend, text, and border to ensure a cohesive and intentionally-designed look.

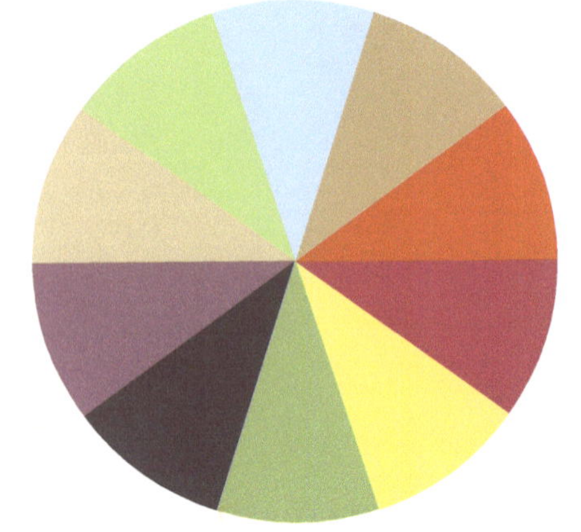

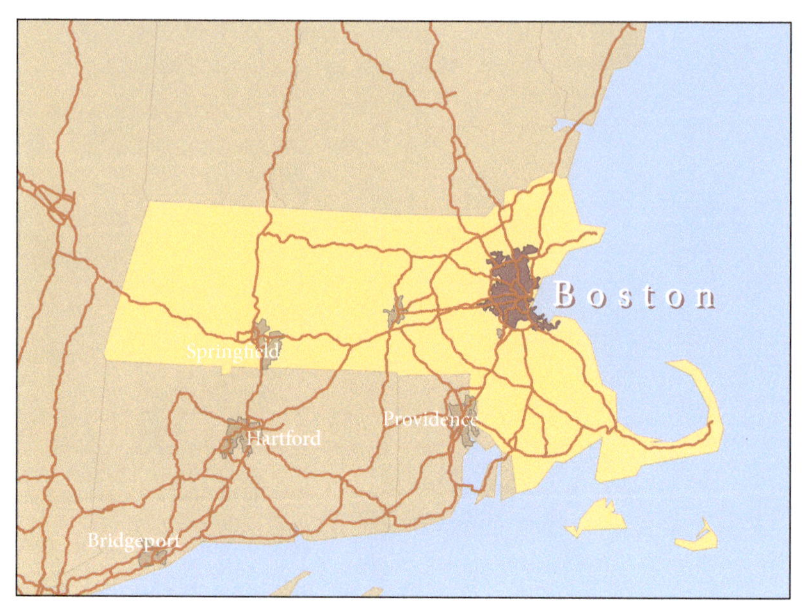

		HEX	RGB	CMYK %
1		9FB480	159 180 128	40 17 60 00
2		5C5252	92 82 82	59 59 56 32
3		9A7683	154 118 131	41 56 37 05
4		DCC8A9	220 200 169	14 19 35 00
5		C3D298	195 210 152	26 07 50 00
6		BED2FF	190 210 255	22 12 00 00
7		C3AA87	195 170 135	25 31 50 00
8		CC6550	204 101 80	16 72 71 03
9		B05266	176 82 102	27 79 47 06
10		F7E68C	247 230 140	04 05 56 00

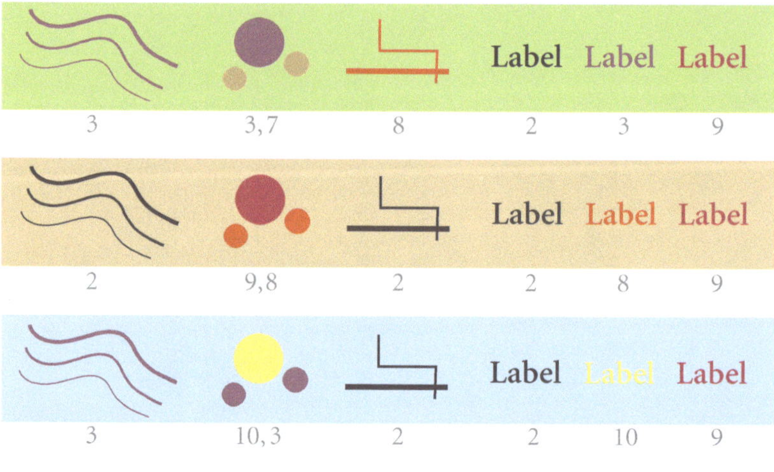

Deuteranope Simulation

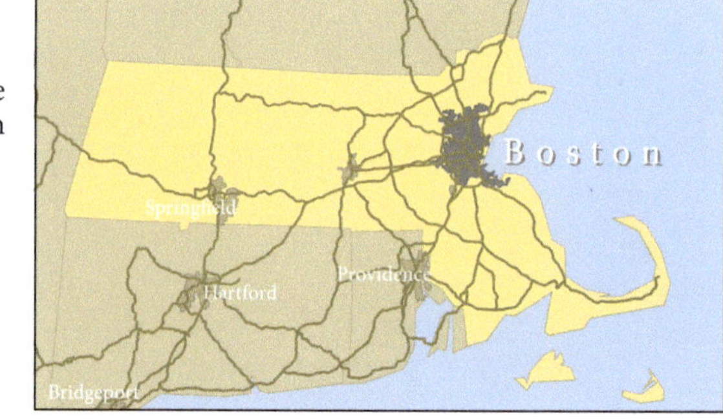

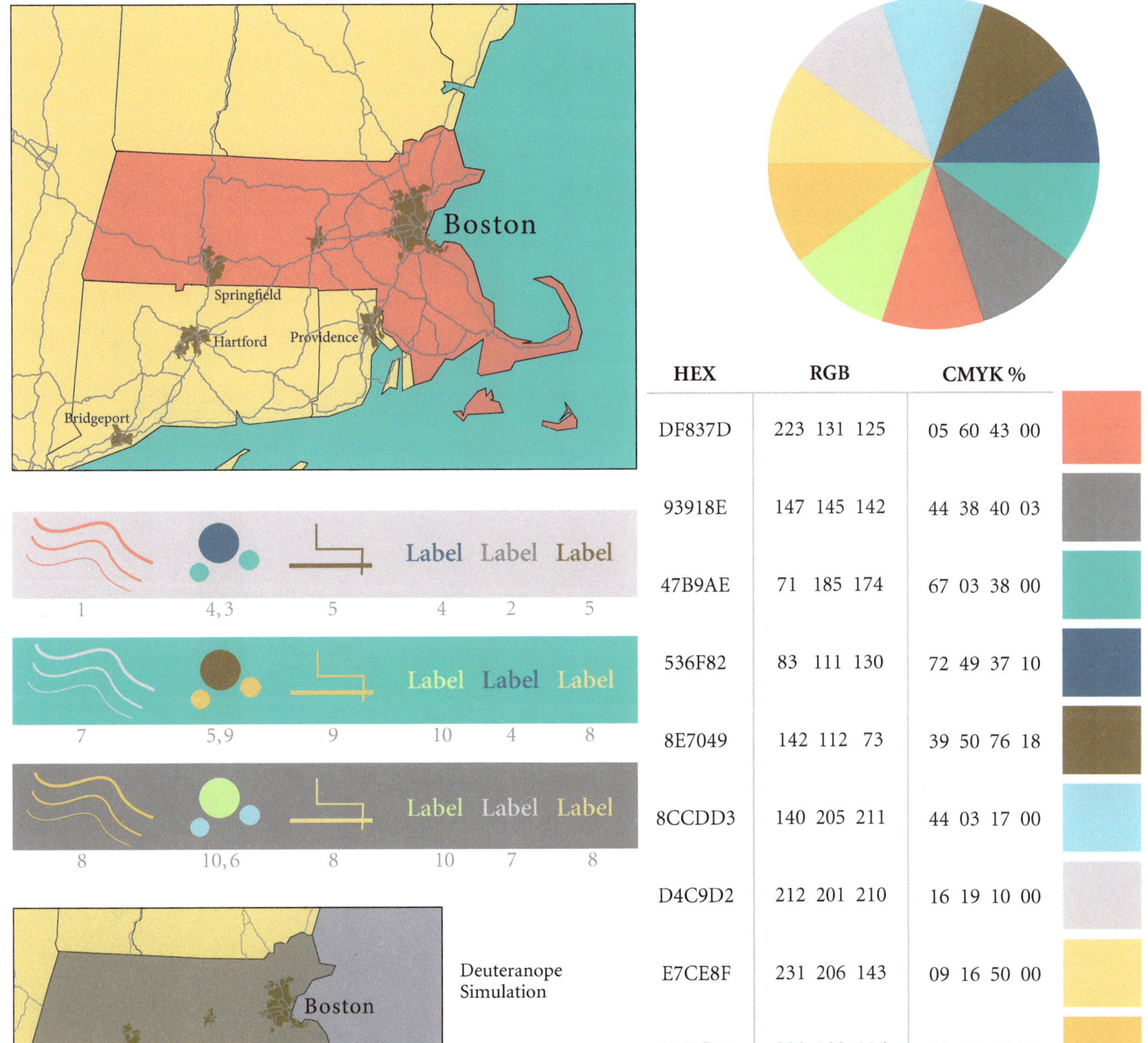

HEX	RGB	CMYK %	
DF837D	223 131 125	05 60 43 00	1
93918E	147 145 142	44 38 40 03	2
47B9AE	71 185 174	67 03 38 00	3
536F82	83 111 130	72 49 37 10	4
8E7049	142 112 73	39 50 76 18	5
8CCDD3	140 205 211	44 03 17 00	6
D4C9D2	212 201 210	16 19 10 00	7
E7CE8F	231 206 143	09 16 50 00	8
E4BC74	228 188 116	11 25 63 00	9
BCDC9D	188 220 157	28 00 49 00	10

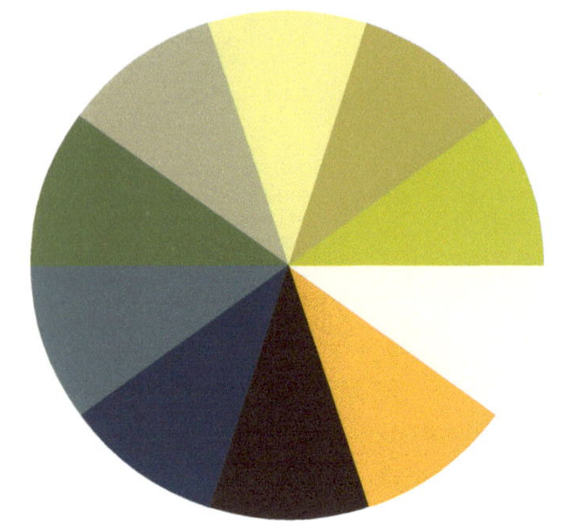

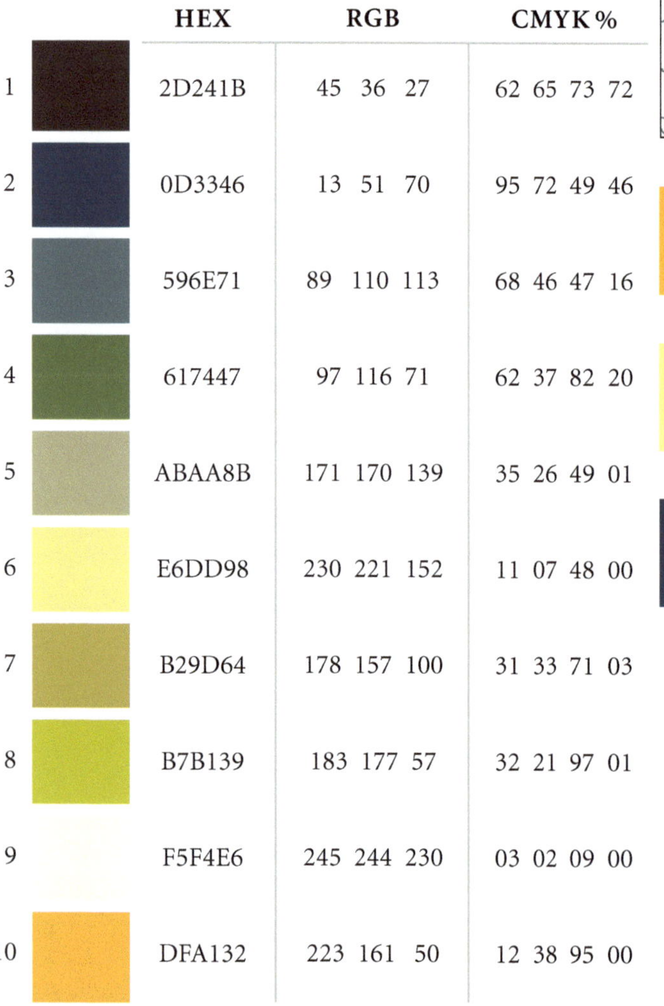

	HEX	RGB	CMYK %
1	2D241B	45 36 27	62 65 73 72
2	0D3346	13 51 70	95 72 49 46
3	596E71	89 110 113	68 46 47 16
4	617447	97 116 71	62 37 82 20
5	ABAA8B	171 170 139	35 26 49 01
6	E6DD98	230 221 152	11 07 48 00
7	B29D64	178 157 100	31 33 71 03
8	B7B139	183 177 57	32 21 97 01
9	F5F4E6	245 244 230	03 02 09 00
10	DFA132	223 161 50	12 38 95 00

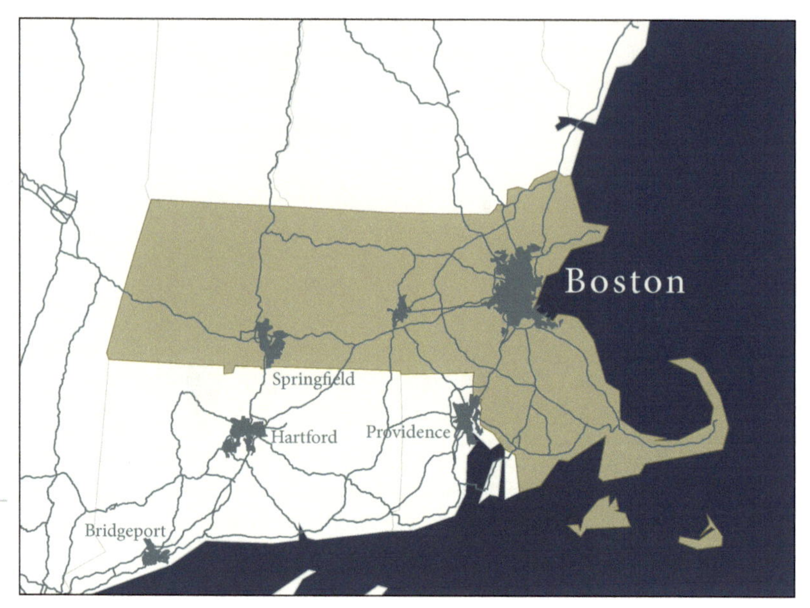

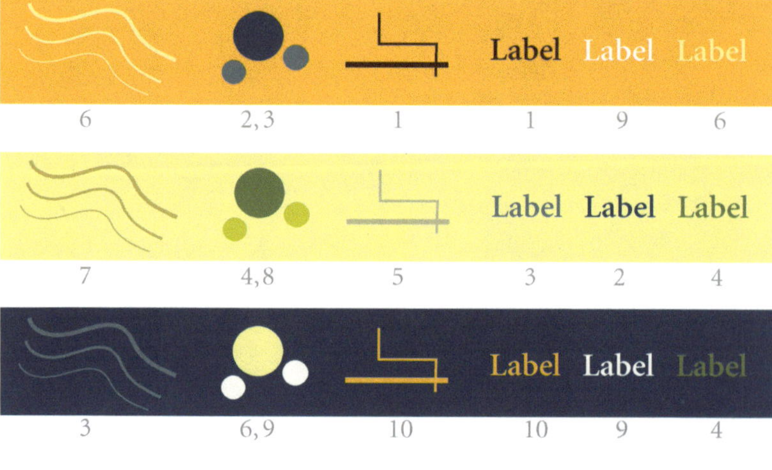

Deuteranope Simulation

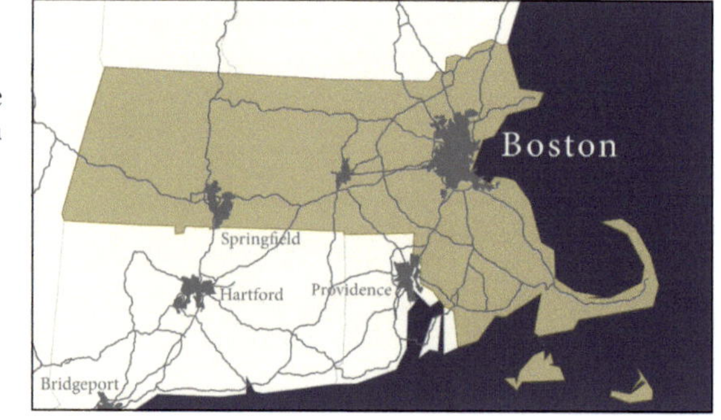

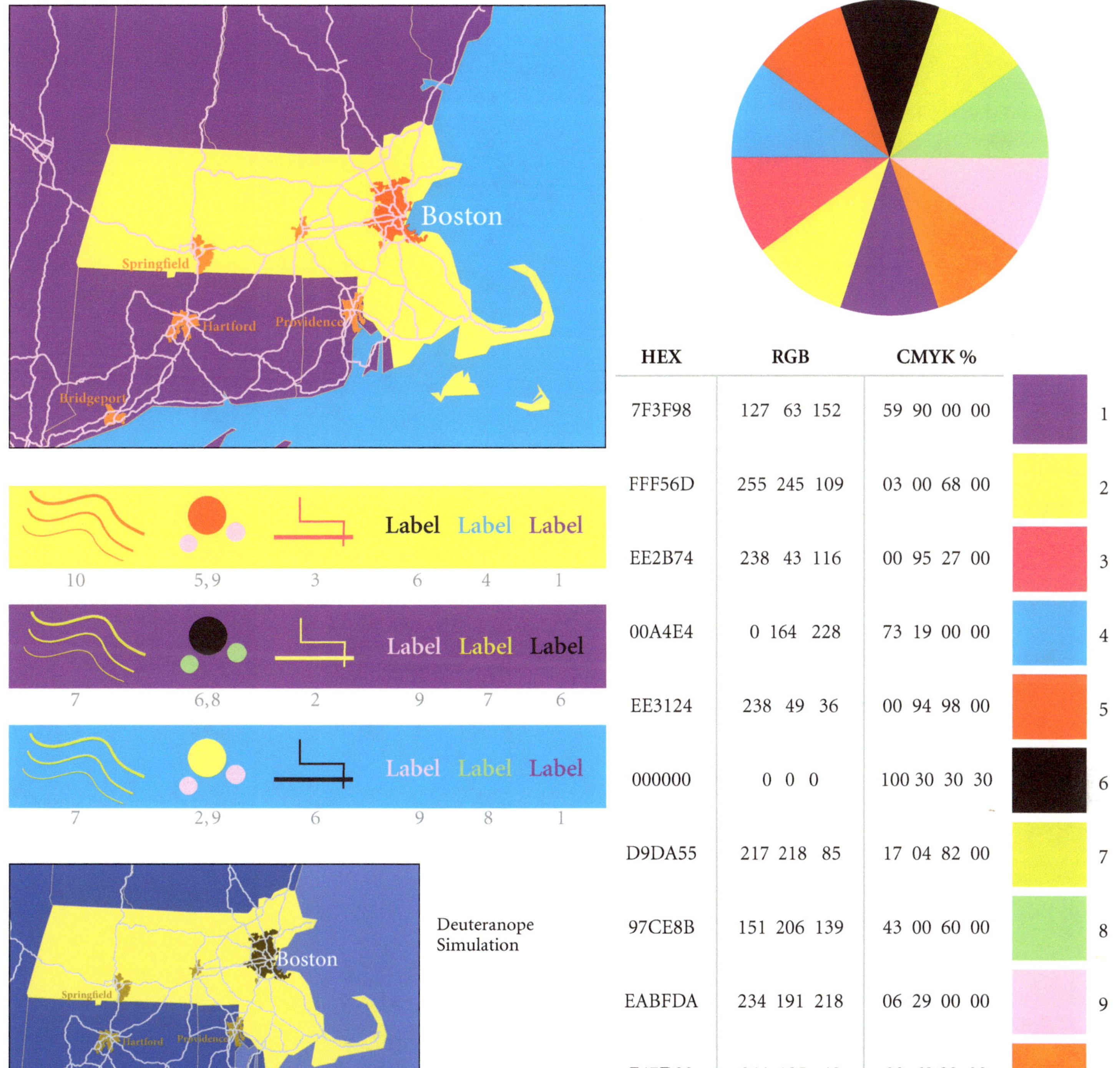

Deuteranope Simulation

HEX	RGB	CMYK %	
7F3F98	127 63 152	59 90 00 00	1
FFF56D	255 245 109	03 00 68 00	2
EE2B74	238 43 116	00 95 27 00	3
00A4E4	0 164 228	73 19 00 00	4
EE3124	238 49 36	00 94 98 00	5
000000	0 0 0	100 30 30 30	6
D9DA55	217 218 85	17 04 82 00	7
97CE8B	151 206 139	43 00 60 00	8
EABFDA	234 191 218	06 29 00 00	9
F47D30	244 125 48	00 63 92 00	10

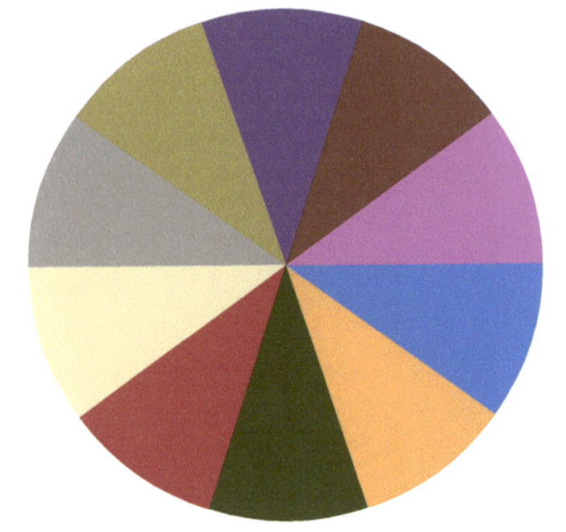

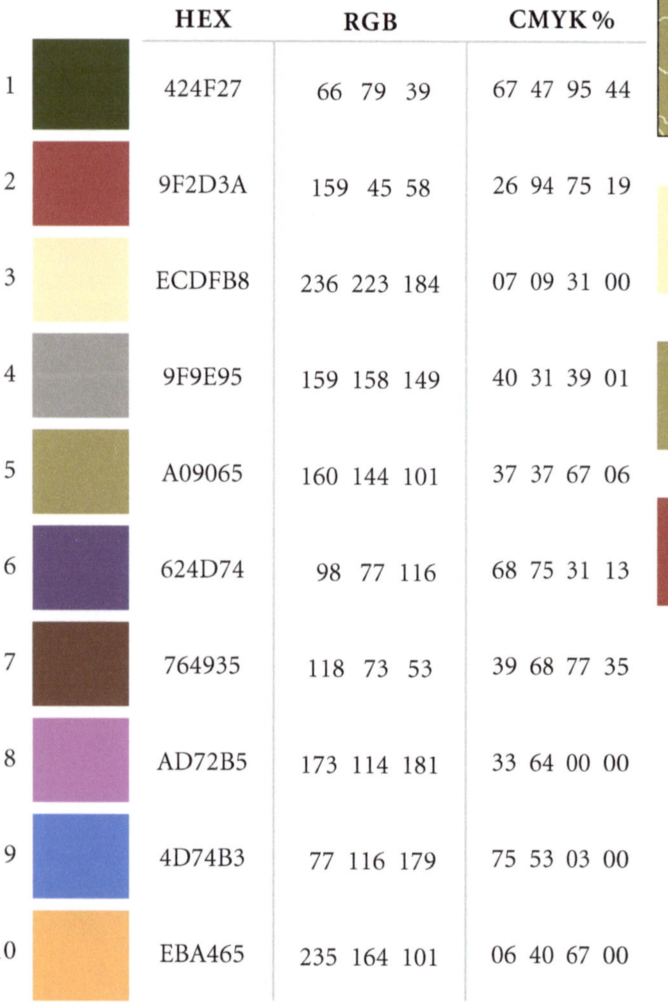

	HEX	RGB	CMYK %
1	424F27	66 79 39	67 47 95 44
2	9F2D3A	159 45 58	26 94 75 19
3	ECDFB8	236 223 184	07 09 31 00
4	9F9E95	159 158 149	40 31 39 01
5	A09065	160 144 101	37 37 67 06
6	624D74	98 77 116	68 75 31 13
7	764935	118 73 53	39 68 77 35
8	AD72B5	173 114 181	33 64 00 00
9	4D74B3	77 116 179	75 53 03 00
10	EBA465	235 164 101	06 40 67 00

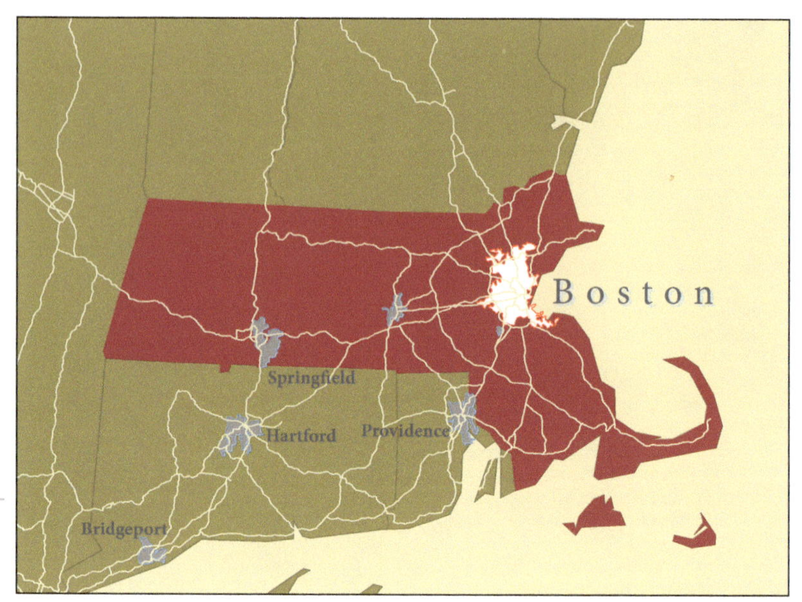

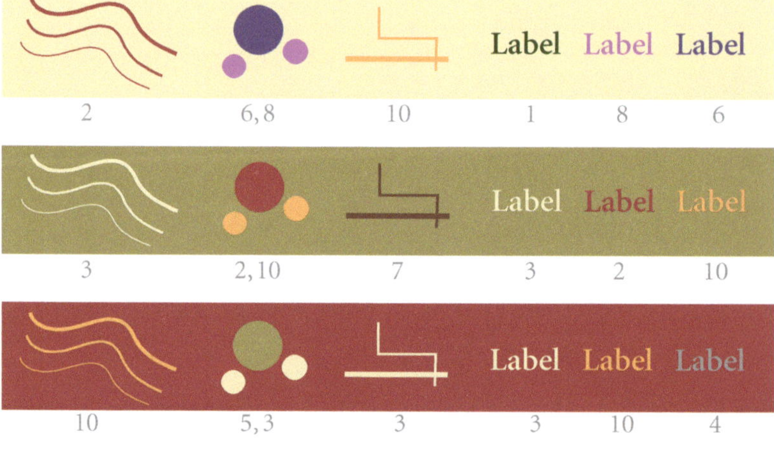

Deuteranope Simulation

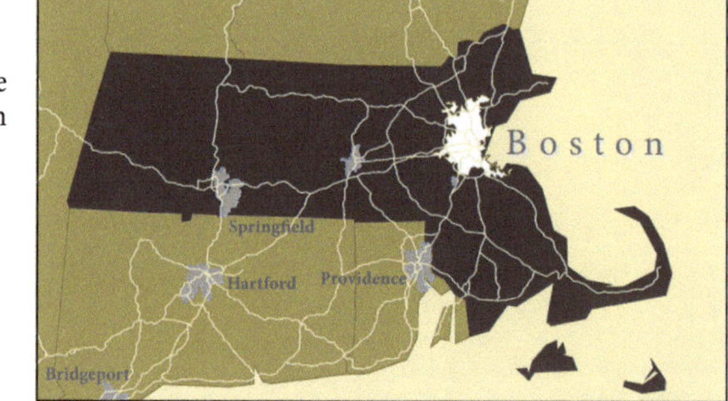

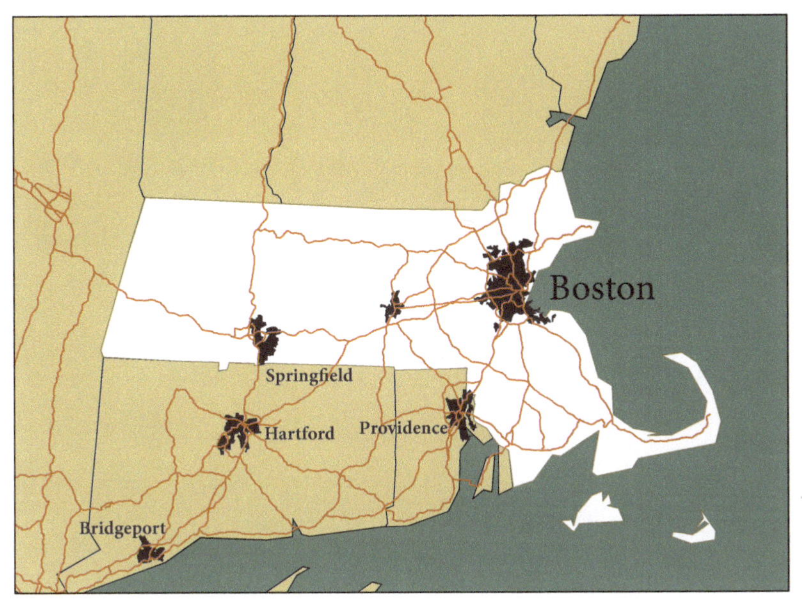

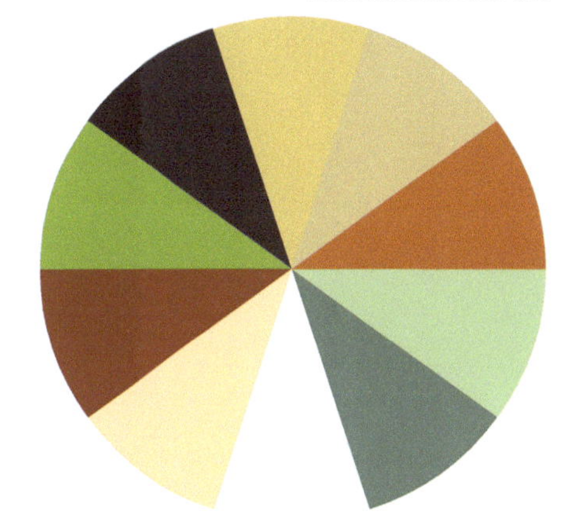

Label Label Label

4 | 9, 4 | 1 | 2 | 10 | 1

Label Label Label

9 | 2, 3 | 9 | 2 | 7 | 9

Label Label Label

5 | 7, 9 | 3 | 1 | 3 | 10

Springfield Hartford Providence Bridgeport Boston

Deuteranope Simulation

HEX	RGB	CMYK %	
FEFAFF	254 250 255	00 02 00 00	1
677D6C	103 125 108	61 38 58 13	2
B3CD9F	179 205 159	32 07 45 00	3
BA5F22	186 95 34	21 71 100 09	4
CDBC8E	205 188 142	21 22 49 00	5
D6C180	214 193 128	47 20 59 00	6
23191A	35 25 26	66 70 65 77	7
8B9E4B	139 158 75	50 24 89 04	8
834021	131 64 33	33 77 95 33	9
FAE2B2	250 226 178	02 10 33 00	10

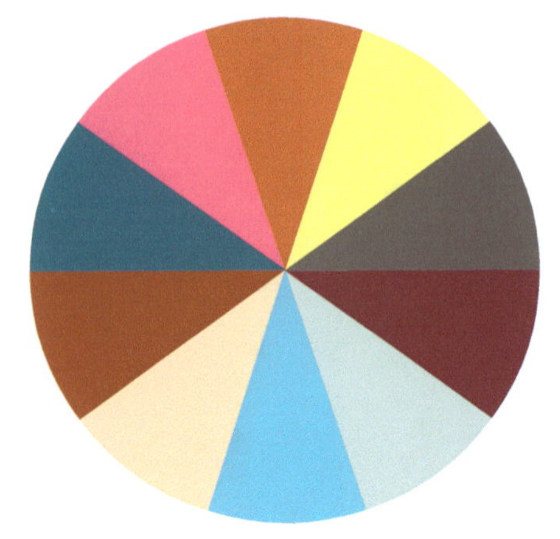

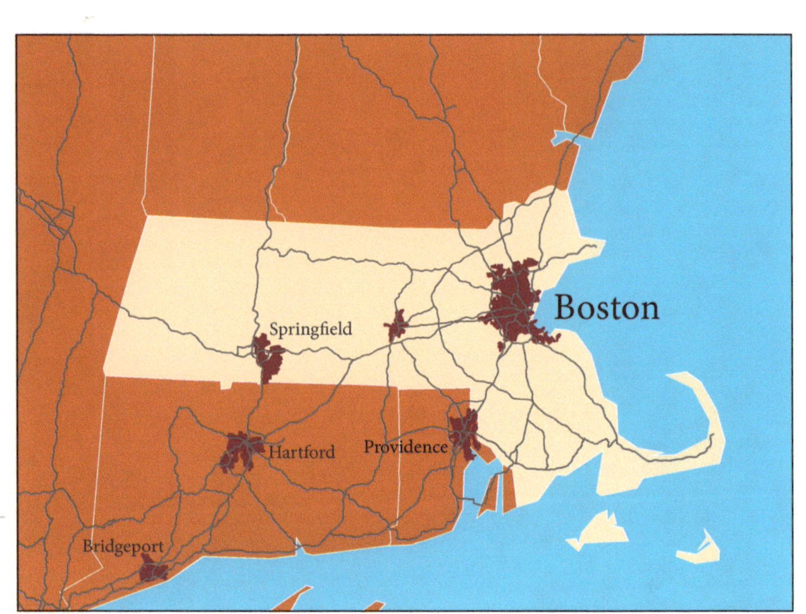

	HEX	RGB	CMYK %
1	4DBAD9	77 186 217	62 06 10 00
2	FFDFBC	255 223 188	00 13 27 00
3	9F4B1D	159 75 29	27 77 100 20
4	036371	3 99 113	91 48 45 18
5	ED669E	237 102 158	01 75 05 00
6	CB552F	203 85 47	15 80 95 04
7	F1FDA3	241 253 163	08 00 44 00
8	776960	119 105 96	51 52 56 20
9	761525	118 21 37	32 98 79 41
10	BAD0D5	186 208 213	27 10 13 00

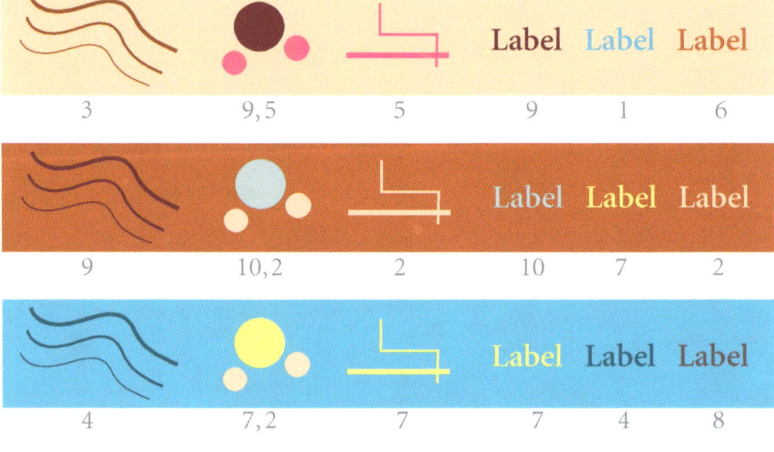

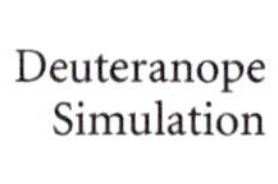

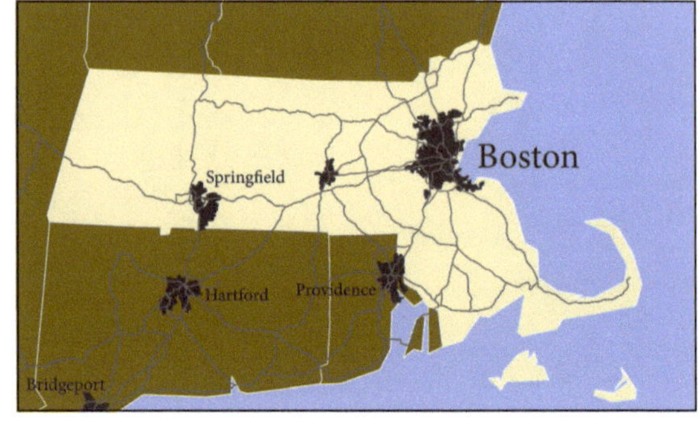

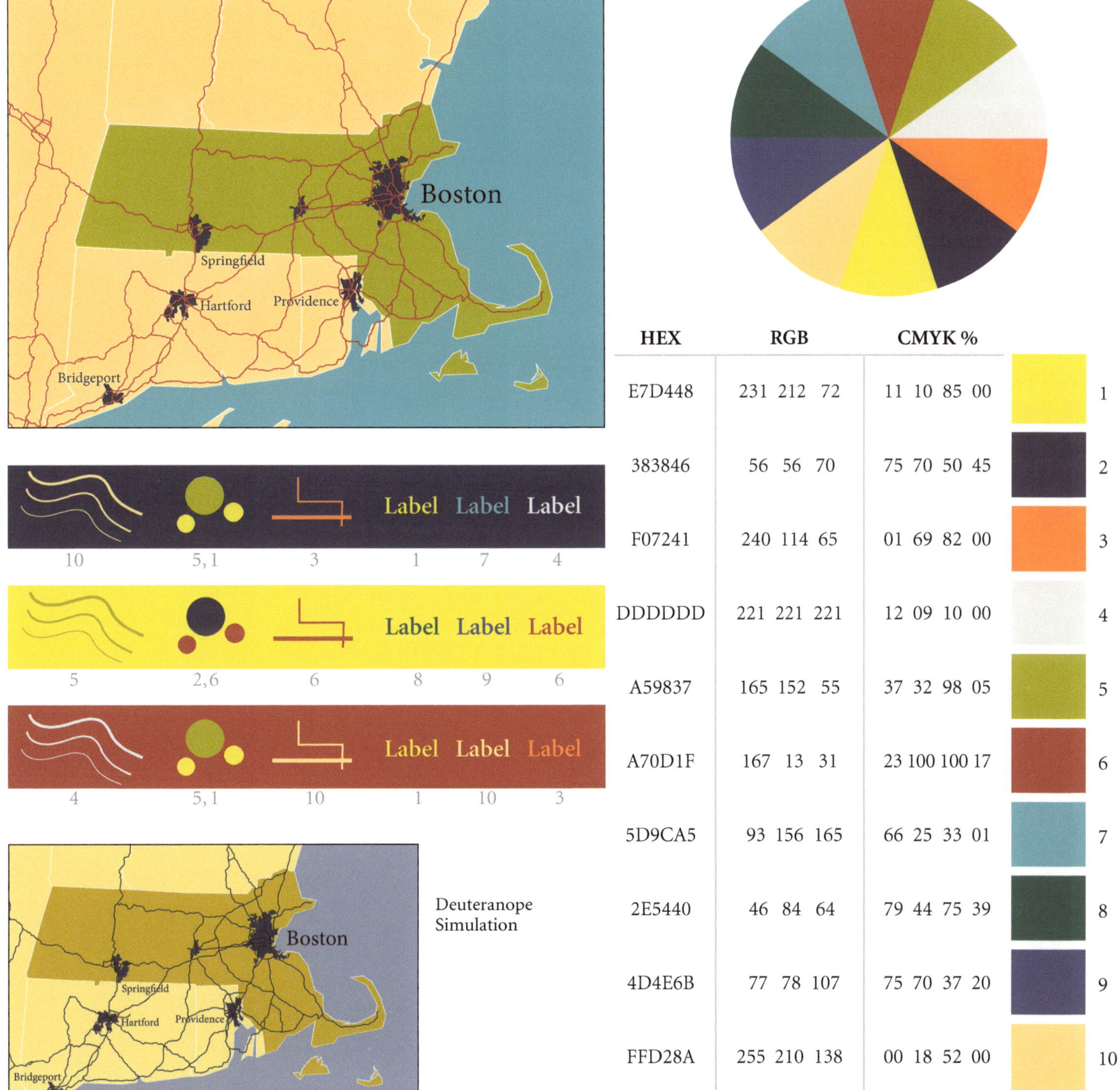

Deuteranope Simulation

HEX	RGB	CMYK %	
E7D448	231 212 72	11 10 85 00	1
383846	56 56 70	75 70 50 45	2
F07241	240 114 65	01 69 82 00	3
DDDDDD	221 221 221	12 09 10 00	4
A59837	165 152 55	37 32 98 05	5
A70D1F	167 13 31	23 100 100 17	6
5D9CA5	93 156 165	66 25 33 01	7
2E5440	46 84 64	79 44 75 39	8
4D4E6B	77 78 107	75 70 37 20	9
FFD28A	255 210 138	00 18 52 00	10

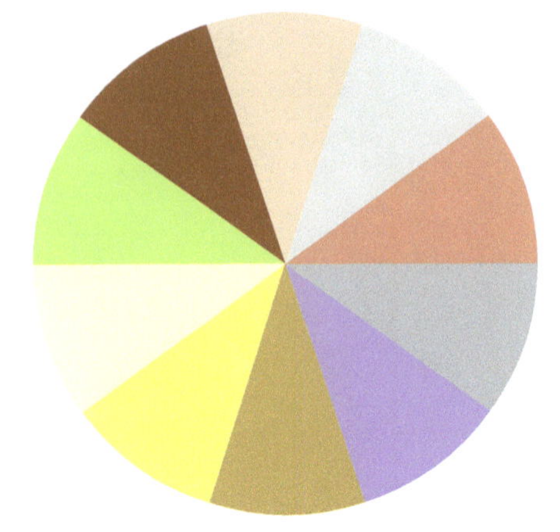

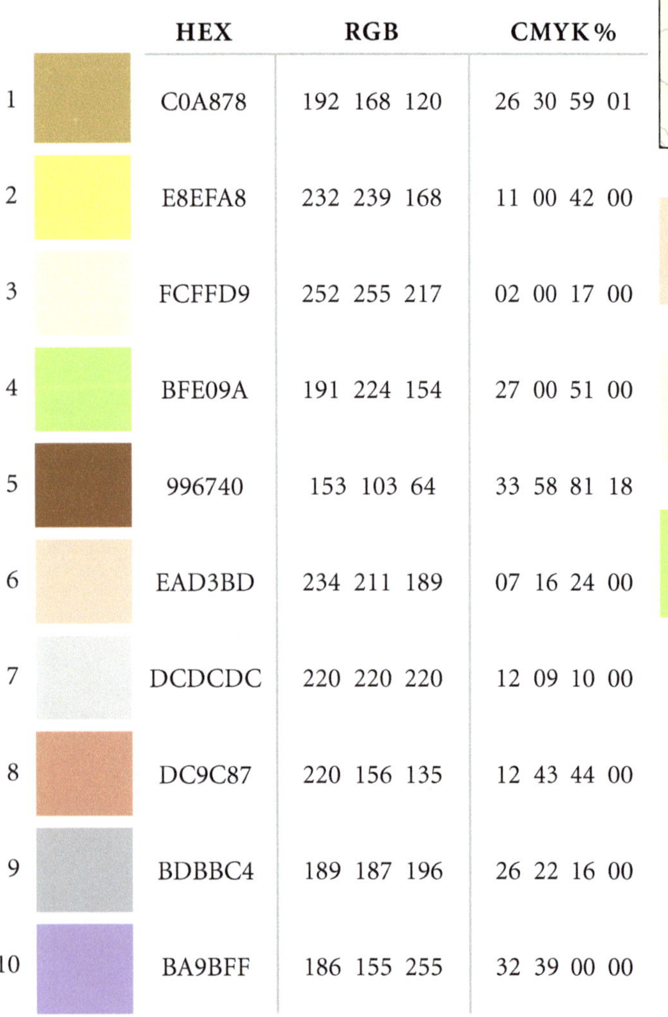

	HEX	RGB	CMYK %
1	C0A878	192 168 120	26 30 59 01
2	E8EFA8	232 239 168	11 00 42 00
3	FCFFD9	252 255 217	02 00 17 00
4	BFE09A	191 224 154	27 00 51 00
5	996740	153 103 64	33 58 81 18
6	EAD3BD	234 211 189	07 16 24 00
7	DCDCDC	220 220 220	12 09 10 00
8	DC9C87	220 156 135	12 43 44 00
9	BDBBC4	189 187 196	26 22 16 00
10	BA9BFF	186 155 255	32 39 00 00

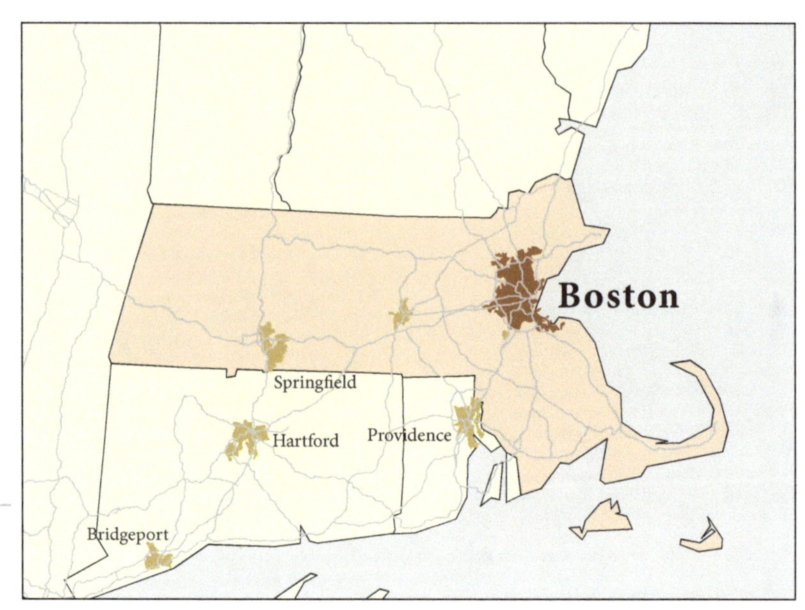

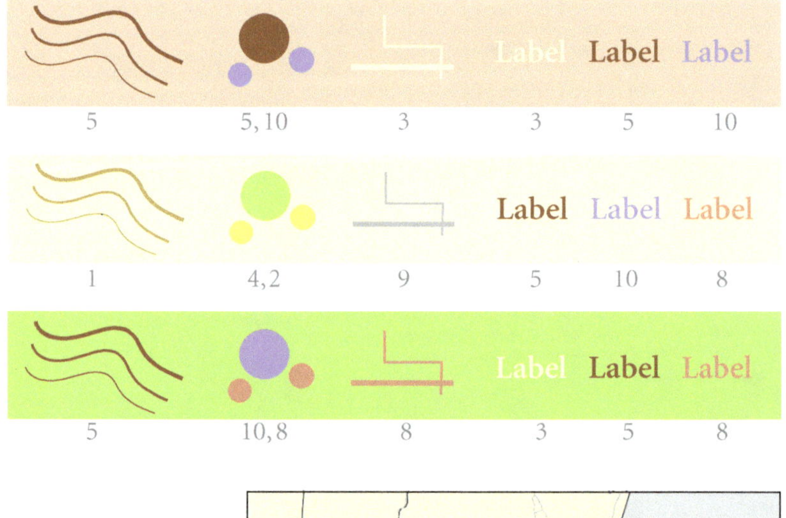

Deuteranope Simulation

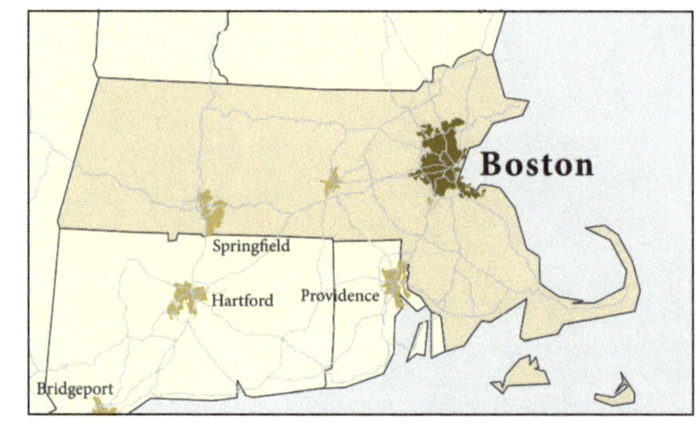

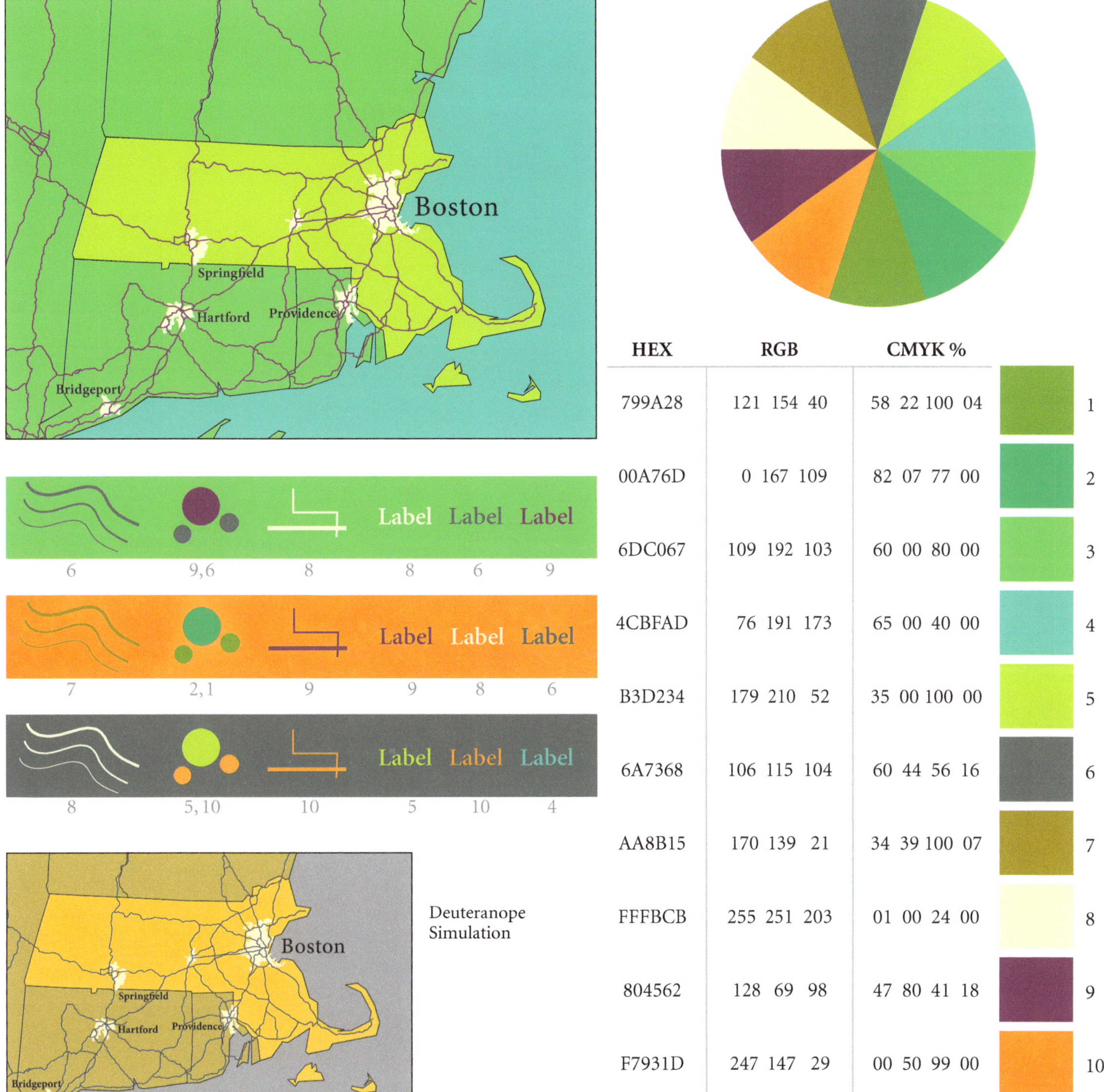

Deuteranope Simulation

HEX	RGB	CMYK %	
799A28	121 154 40	58 22 100 04	1
00A76D	0 167 109	82 07 77 00	2
6DC067	109 192 103	60 00 80 00	3
4CBFAD	76 191 173	65 00 40 00	4
B3D234	179 210 52	35 00 100 00	5
6A7368	106 115 104	60 44 56 16	6
AA8B15	170 139 21	34 39 100 07	7
FFFBCB	255 251 203	01 00 24 00	8
804562	128 69 98	47 80 41 18	9
F7931D	247 147 29	00 50 99 00	10

DIFFERENTIATED PALETTES

Some maps require the use of colors that are easily distinguishable from one another. This differentiation aids in the identification of features by making it easy to match colors from the map with the map key. These palettes do just that while also striving for some color harmony.

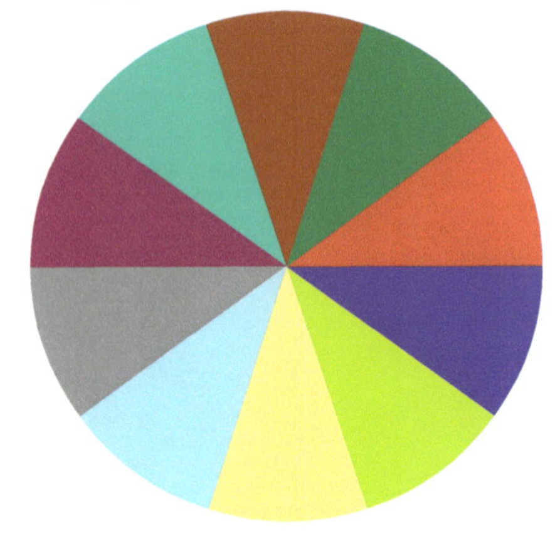

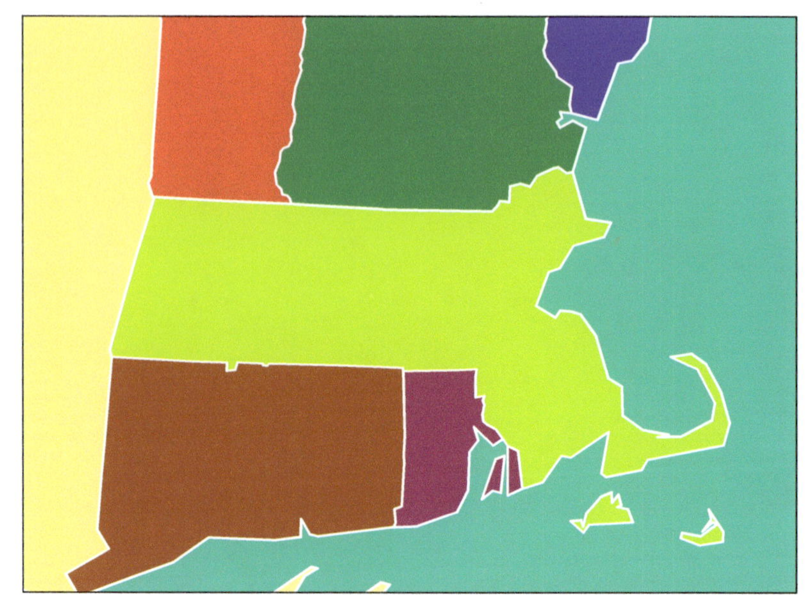

	HEX	RGB	CMYK %
1	FFE791	255 231 145	02 07 52 00
2	96E1EB	150 225 235	37 00 09 00
3	919693	145 150 147	46 34 38 02
4	8E0554	142 5 84	38 100 41 19
5	32B29D	50 178 157	73 05 48 00
6	9B4F19	155 79 25	28 74 100 21
7	007A3D	0 122 61	89 27 100 15
8	D81E05	216 30 5	09 99 100 02
9	46158F	70 21 143	89 100 04 02
10	ABF943	171 249 67	35 00 98 00

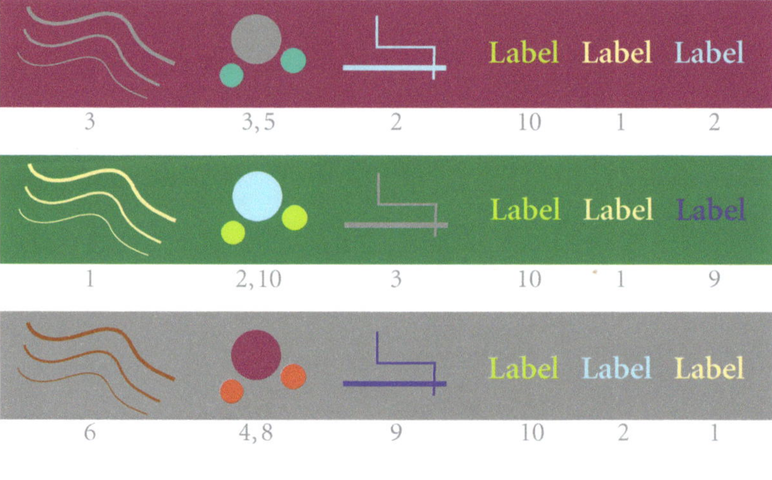

Deuteranope Simulation

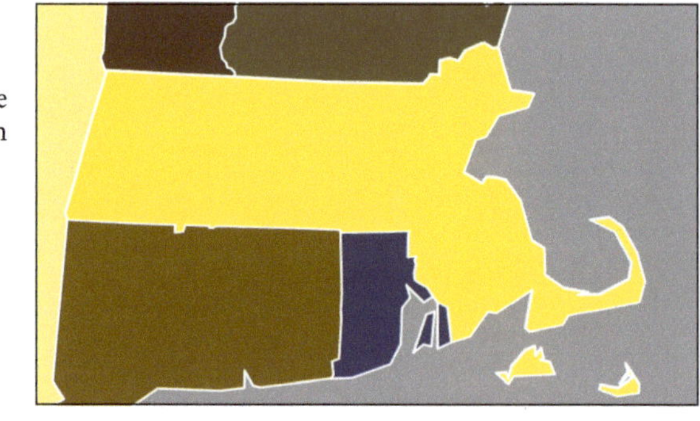

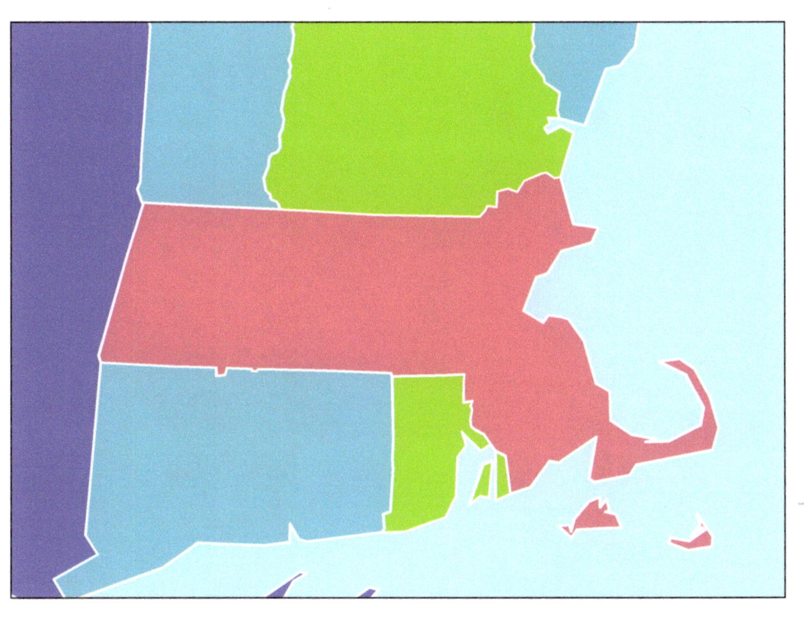

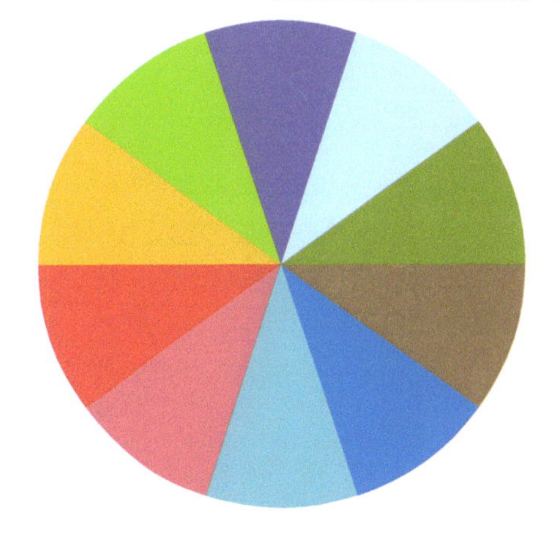

Label Label Label

4 8, 5 2 4 7 5

Label Label Label

4 6, 1 5 7 5 4

Label Label Label

7 9, 4 4 7 10 9

Deuteranope Simulation

HEX	RGB	CMYK %	
75BACA	117 186 202	53 10 17 00	1
DA7A92	218 122 146	12 64 24 00	2
DF014A	223 1 74	06 100 65 01	3
E3B62E	227 182 46	12 27 96 00	4
94D82D	148 216 45	46 00 100 00	5
746DA3	116 109 163	62 61 11 00	6
B6F0FC	182 240 252	25 00 03 00	7
869747	134 151 71	51 27 91 06	8
9E7B58	158 123 88	35 48 70 11	9
1C83EA	28 131 234	76 46 00 00	10

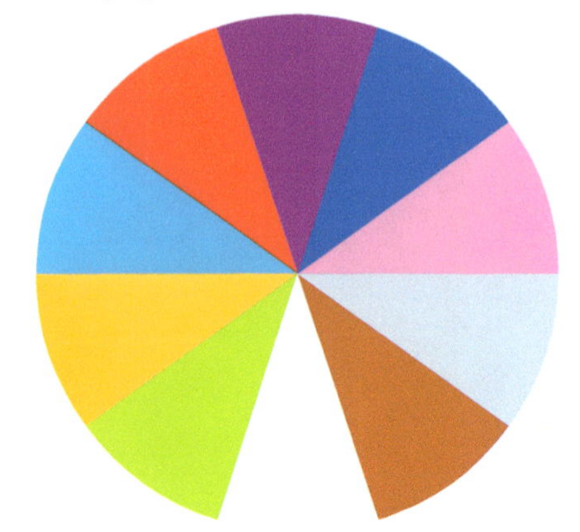

	HEX	RGB	CMYK %
1	F8F8F8	248 248 248	02 01 01 00
2	B3D334	179 211 52	35 00 99 00
3	FFC40A	255 196 10	00 24 99 00
4	00ABF0	0 171 240	69 17 00 00
5	EF1C25	239 28 37	00 99 96 01
6	87318C	135 49 140	56 97 05 00
7	0160A0	1 96 160	95 65 10 01
8	F59EE5	245 158 229	08 43 00 00
9	C3D6E4	195 214 228	22 08 05 00
10	CA6203	202 98 3	16 71 100 04

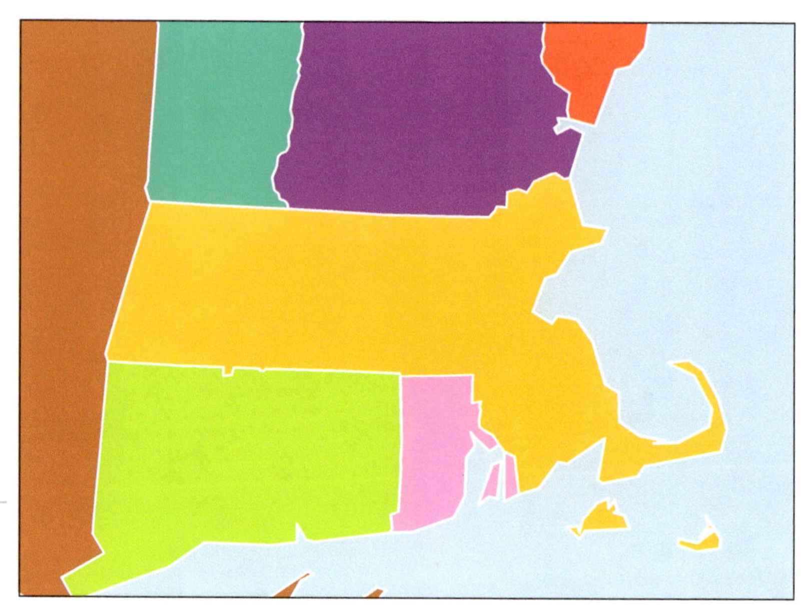

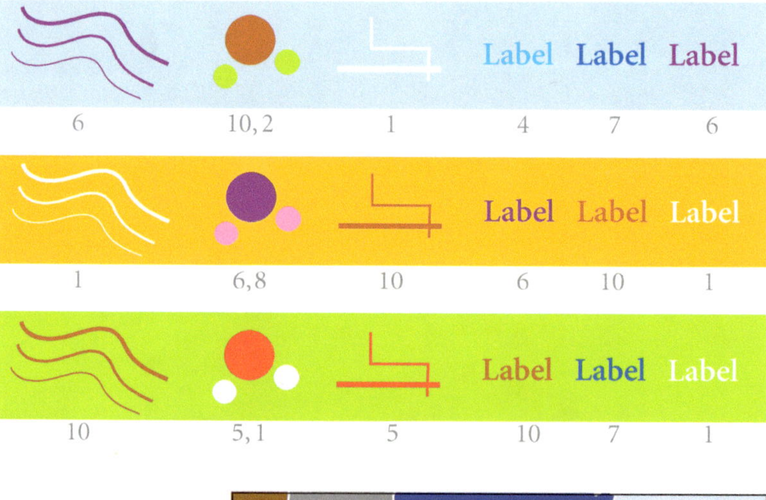

Deuteranope Simulation

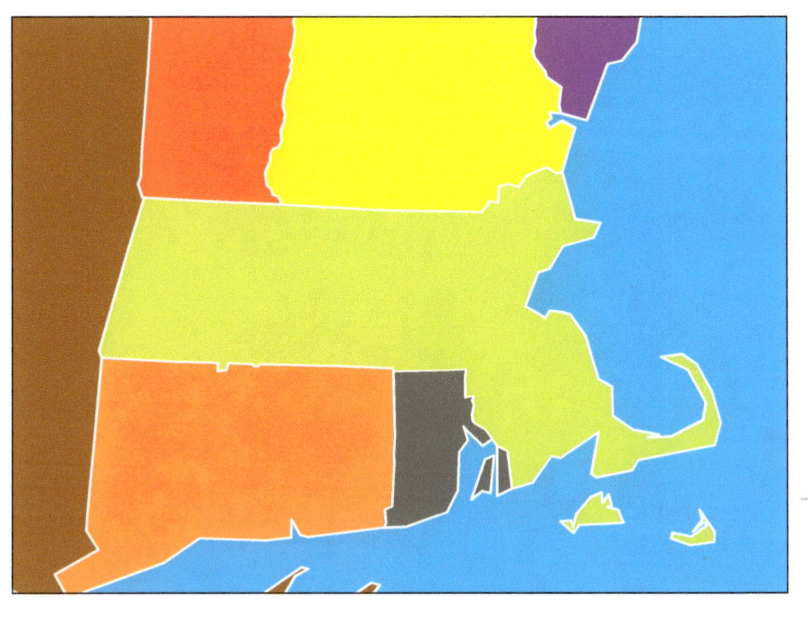

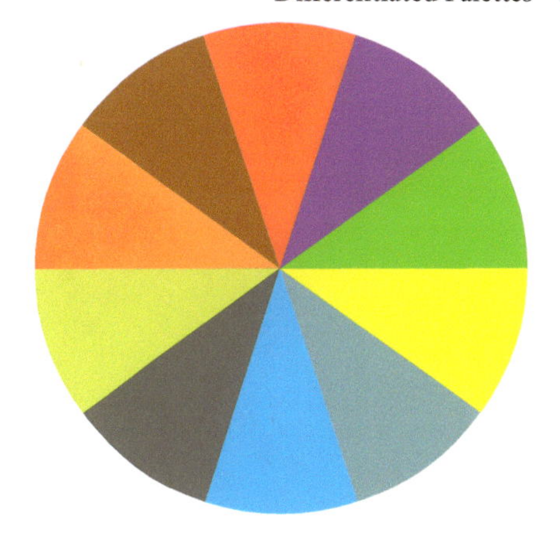

Label Label Label

3 8, 3 6 3 9 4

Label Label Label

5 1, 10 6 7 6 1

Label Label Label

5 7, 3 2 2 7 9

Deuteranope Simulation

HEX	RGB	CMYK %	
06A2E2	6 162 226	73 20 00 00	1
7B7464	123 116 100	50 46 59 16	2
CDD768	205 215 104	22 03 74 00	3
FF8B3F	255 139 63	00 56 82 00	4
A4651E	164 101 30	28 62 100 15	5
FE5A3C	254 90 60	00 80 80 00	6
865793	134 87 147	54 76 13 01	7
75B63F	117 182 63	60 06 100 01	8
F4F134	244 240 52	08 00 88 00	9
79A4A7	121 164 167	55 24 32 00	10

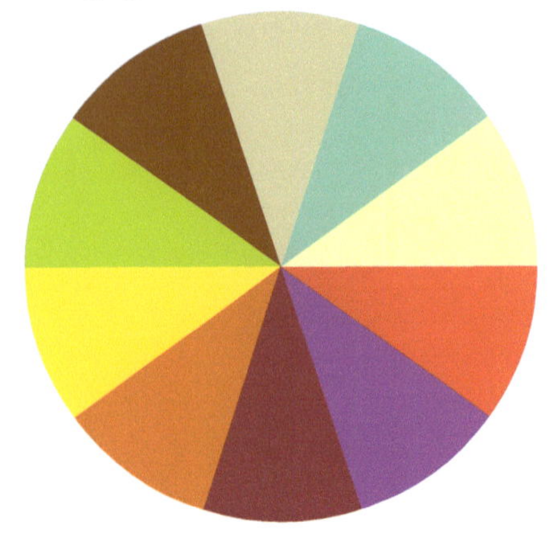

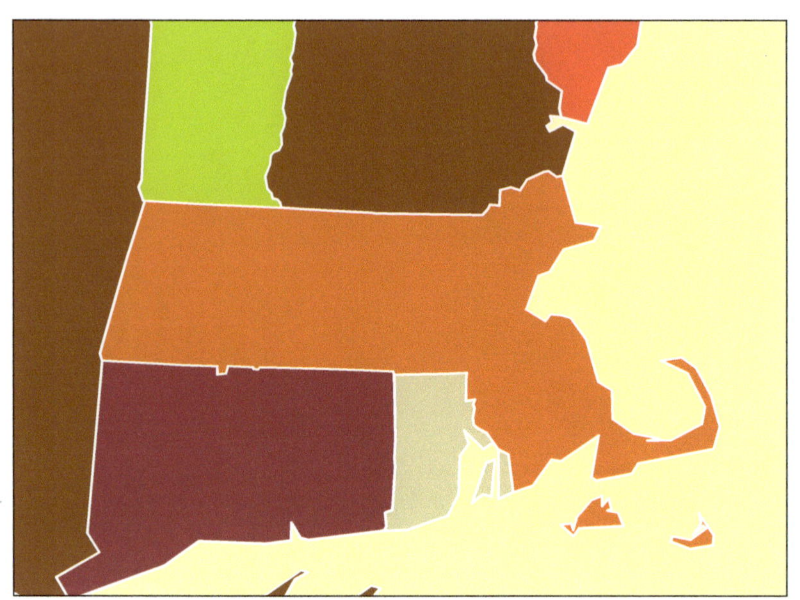

		HEX	RGB	CMYK %
1		821F29	130 31 41	30 96 81 34
2		D46600	212 102 0	13 71 100 02
3		F6E03F	246 224 63	06 06 87 00
4		ADBD06	173 189 6	38 11 100 00
5		7D4A04	125 74 4	37 67 100 34
6		CFC8A2	207 200 162	20 16 40 00
7		80BCA3	128 188 163	51 08 42 00
8		F6F7BD	246 247 189	04 00 32 00
9		DE3F33	222 63 51	07 90 89 01
10		895793	137 87 147	53 77 13 00

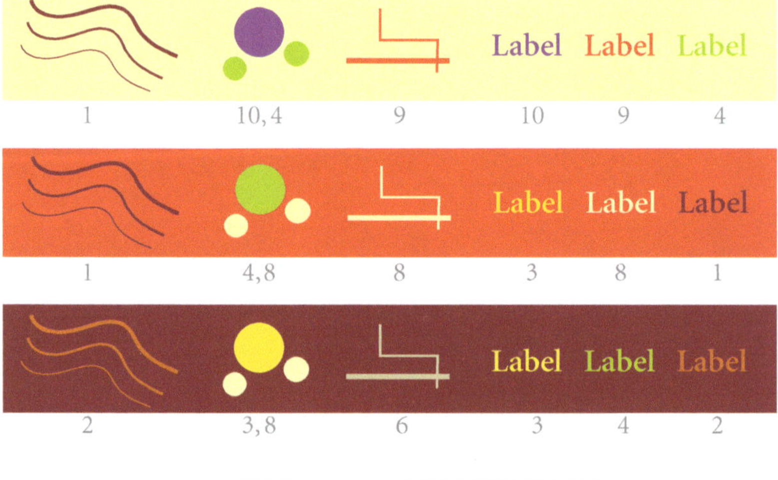

Deuteranope Simulation

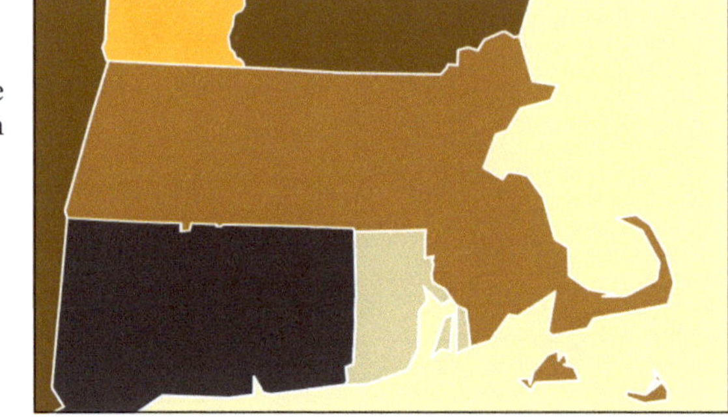

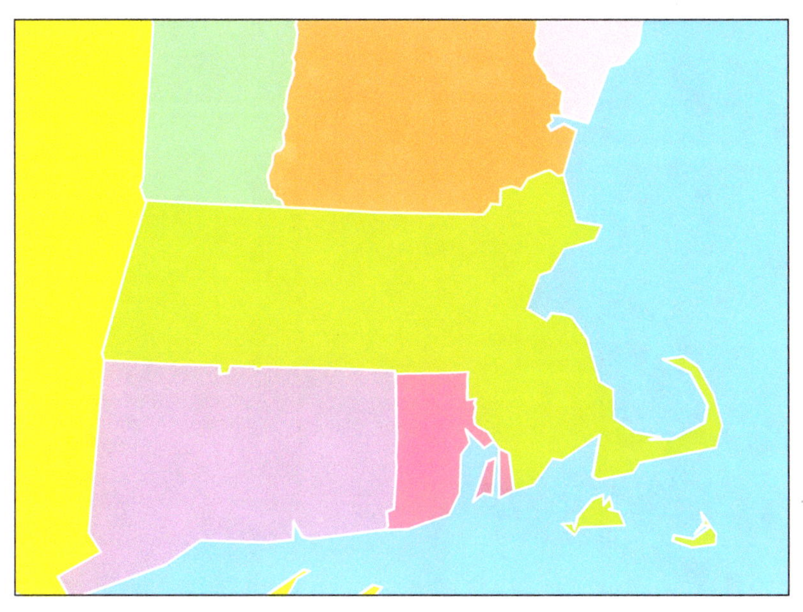

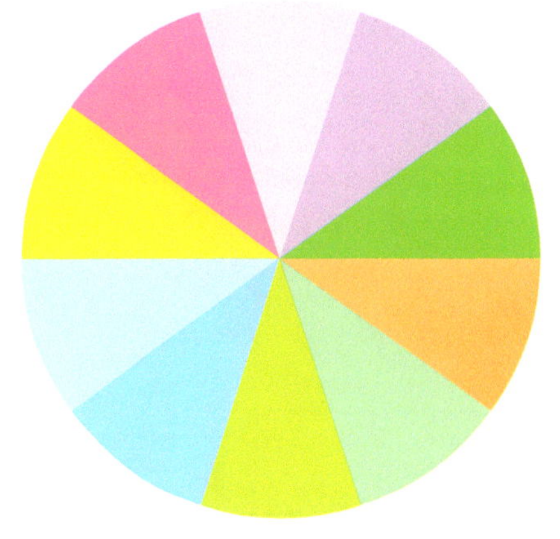

Label Label Label

9 8, 1 9 8 5 1

Label Label Label

7 5, 6 5 6 5 8

Label Label Label

4 7, 6 6 4 7 6

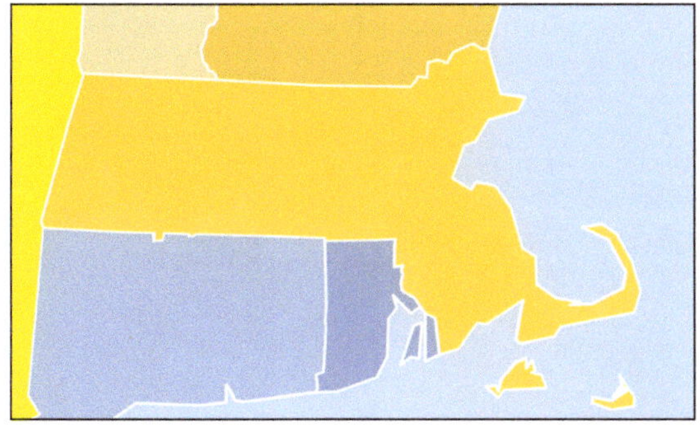

Deuteranope Simulation

HEX	RGB	CMYK %	
CEE451	206 228 81	23 00 83 00	1
99DAF6	153 218 246	36 00 01 00	2
C5EAFC	197 234 252	20 00 00 00	3
FFF210	255 242 16	04 00 92 00	4
FF97CB	255 151 203	00 51 00 00	5
FEE3F6	254 227 246	00 13 00 00	6
D3B3D8	211 179 216	15 31 00 00	7
82D81A	130 216 26	52 00 100 00	8
F9BE64	249 190 100	02 27 70 00	9
B4E0AD	180 224 173	30 00 41 00	10

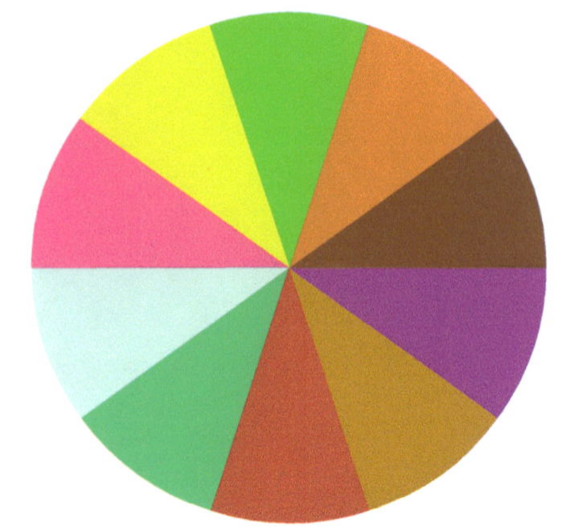

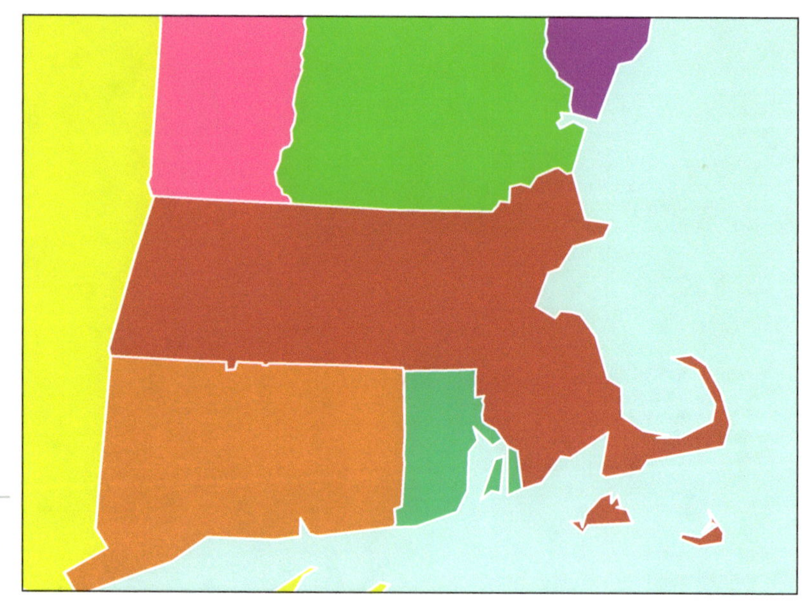

		HEX	RGB	CMYK %
1		C23231	194 50 49	17 94 90 06
2		1FB079	31 176 121	77 03 71 00
3		B1DED7	177 222 215	30 00 17 00
4		DF56B0	223 86 176	13 79 00 00
5		D1DA1B	209 218 27	22 01 100 01
6		55C034	85 192 52	67 00 100 00
7		DA7932	218 121 50	12 62 94 01
8		875532	135 85 50	35 65 86 27
9		8E319B	142 49 155	53 95 00 00
10		BA831F	186 131 31	25 48 100 06

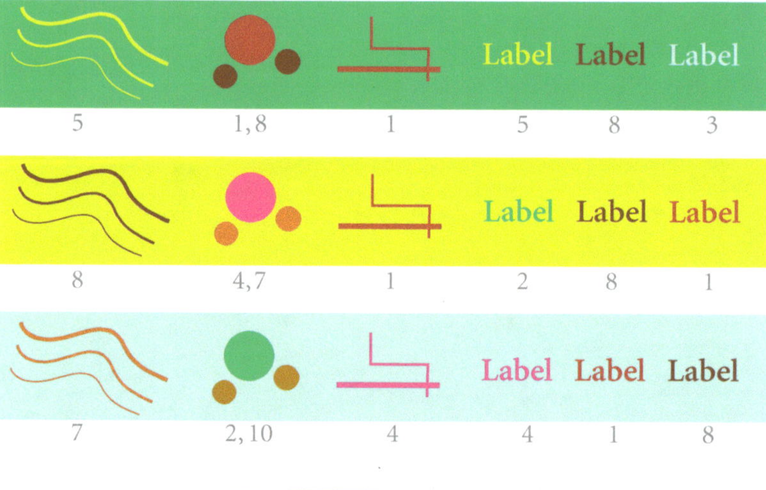

Deuteranope Simulation

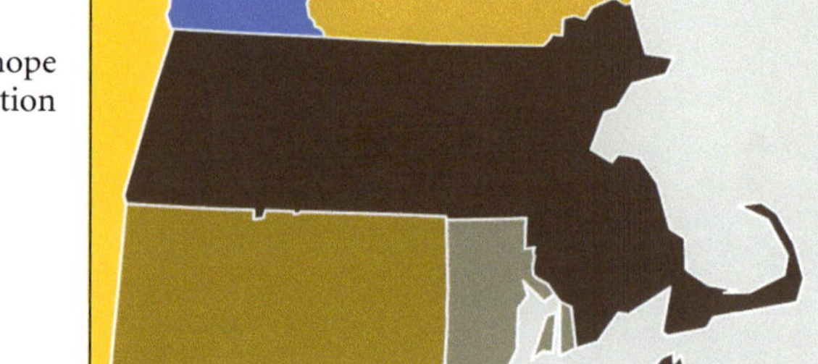

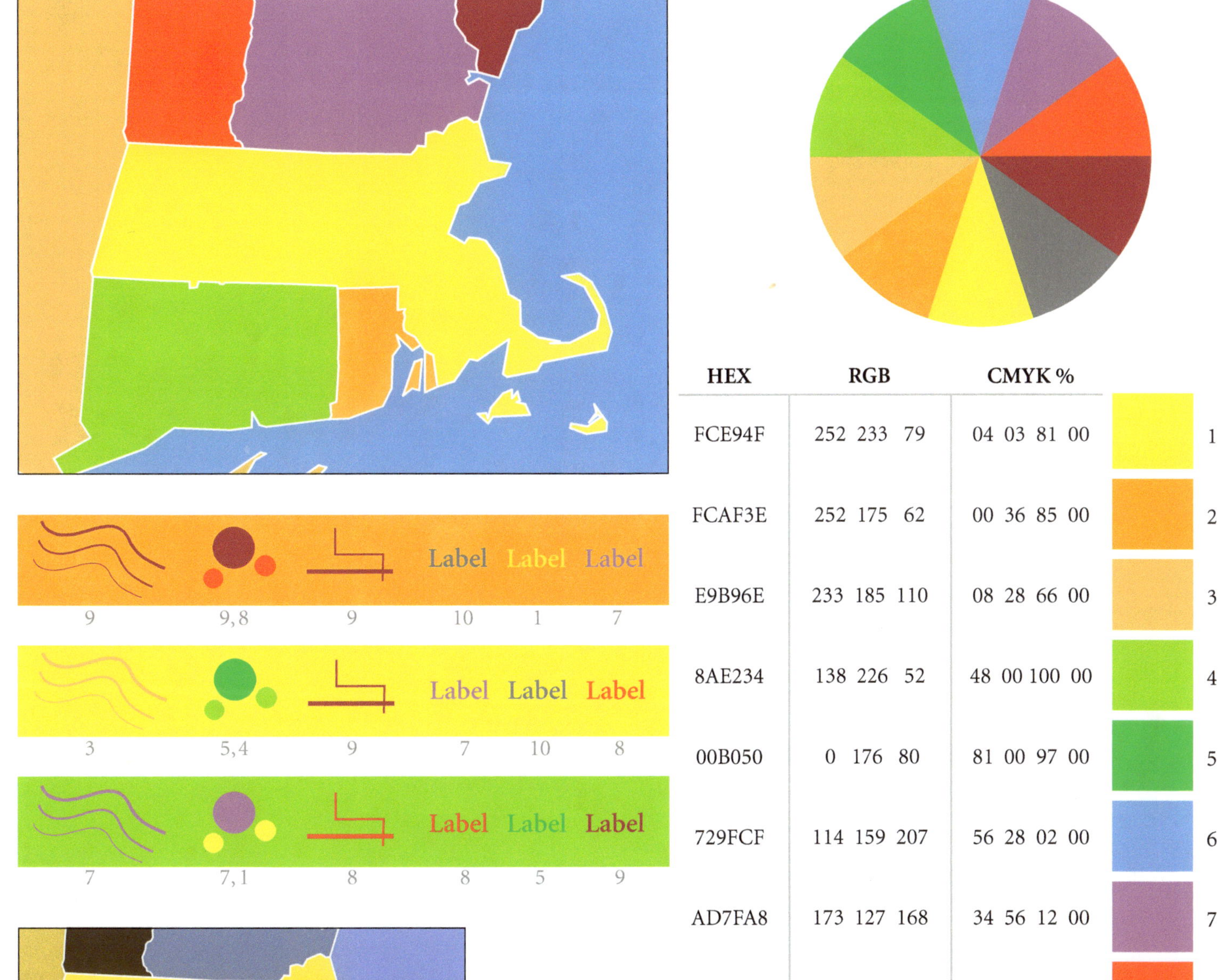

HEX	RGB	CMYK %	
FCE94F	252 233 79	04 03 81 00	1
FCAF3E	252 175 62	00 36 85 00	2
E9B96E	233 185 110	08 28 66 00	3
8AE234	138 226 52	48 00 100 00	4
00B050	0 176 80	81 00 97 00	5
729FCF	114 159 207	56 28 02 00	6
AD7FA8	173 127 168	34 56 12 00	7
EF2929	239 41 41	00 96 94 00	8
A40033	164 0 51	24 100 77 18	9
888A85	136 138 133	49 39 44 05	10

Deuteranope Simulation

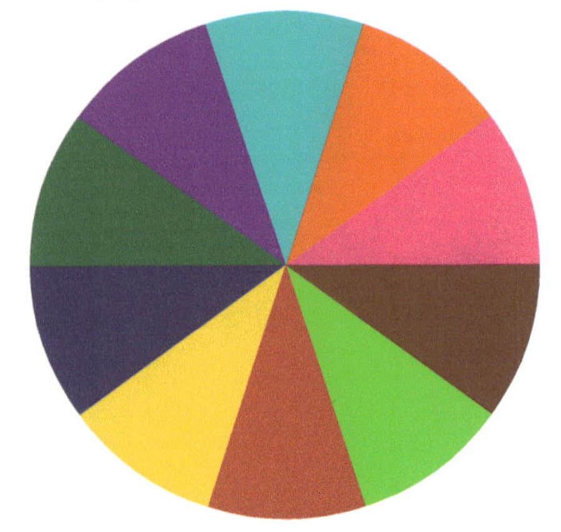

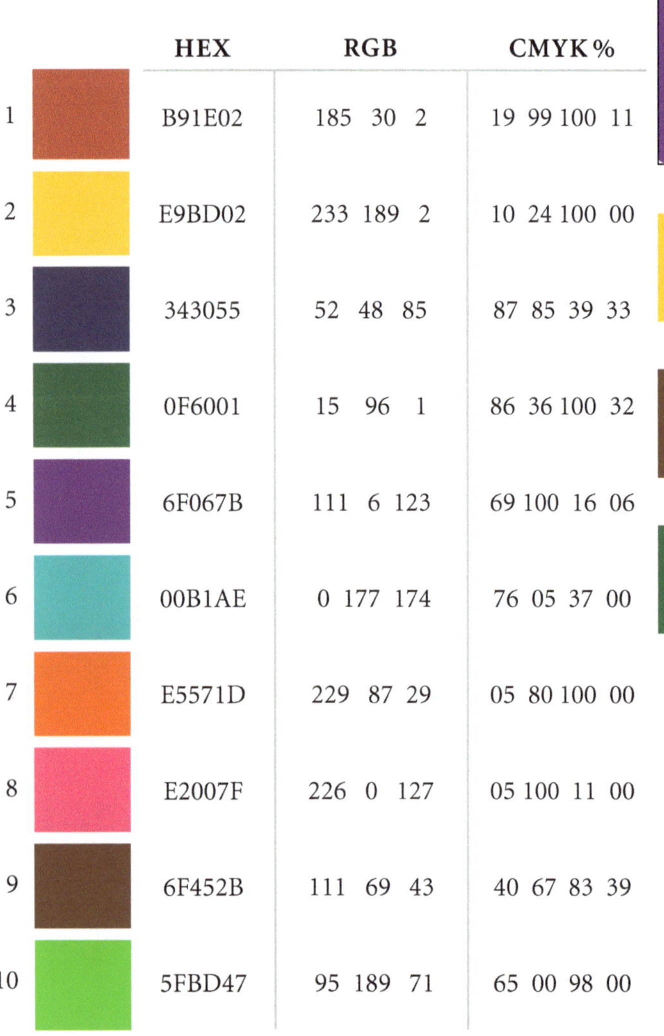

	HEX	RGB	CMYK %
1	B91E02	185 30 2	19 99 100 11
2	E9BD02	233 189 2	10 24 100 00
3	343055	52 48 85	87 85 39 33
4	0F6001	15 96 1	86 36 100 32
5	6F067B	111 6 123	69 100 16 06
6	00B1AE	0 177 174	76 05 37 00
7	E5571D	229 87 29	05 80 100 00
8	E2007F	226 0 127	05 100 11 00
9	6F452B	111 69 43	40 67 83 39
10	5FBD47	95 189 71	65 00 98 00

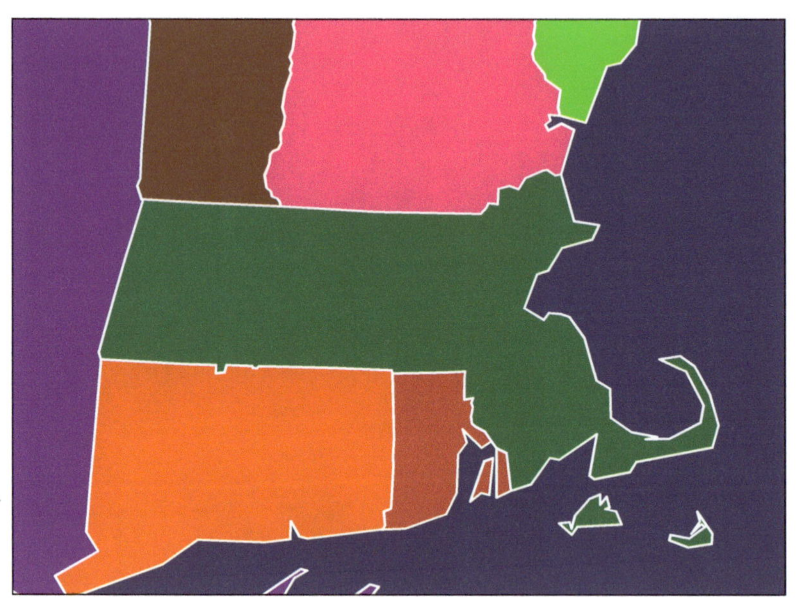

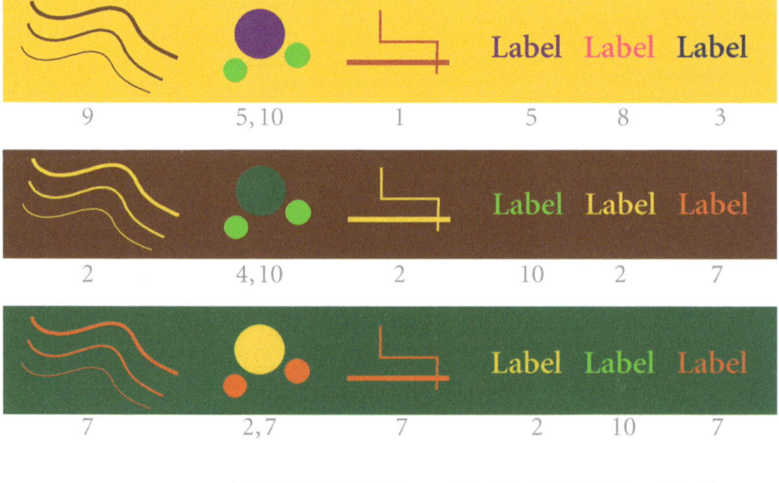

Deuteranope Simulation

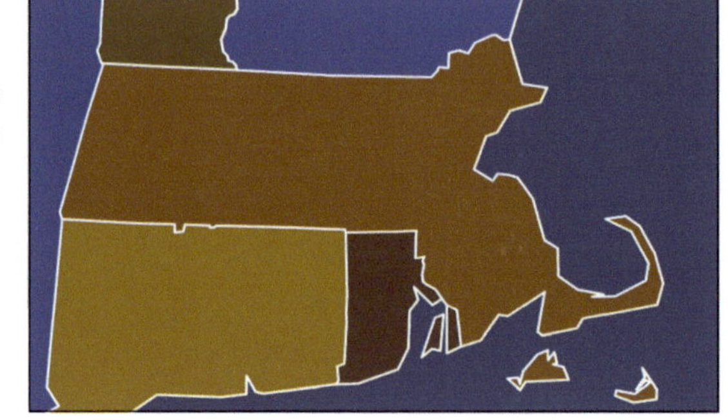

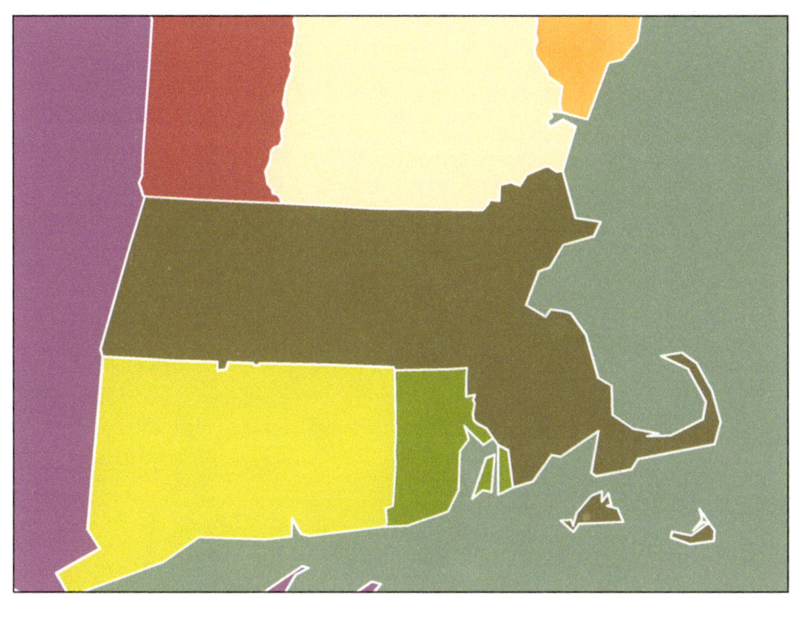

Label Label Label

7 10, 1 5 5 10 1

Label Label Label

8 3, 6 6 8 9 4

Label Label Label

5 4, 1 10 10 1 5

Deuteranope Simulation

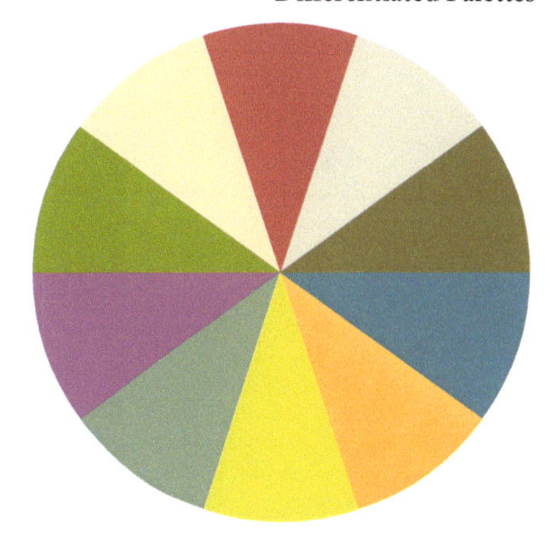

HEX	RGB	CMYK %	
DFCD11	223 205 17	16 12 100 00	1
889E7F	136 158 127	50 26 56 03	2
A06C90	160 108 144	40 65 23 01	3
8C9208	140 146 8	49 30 100 07	4
FAE9BD	250 233 189	02 07 29 00	5
B94A51	185 74 81	21 83 64 07	6
E1DAC5	225 218 197	11 10 22 00	7
857548	133 117 72	44 44 79 18	8
567C87	86 124 135	70 41 39 07	9
FFB752	255 183 82	00 32 77 00	10

COLOR RAMPS

Color ramps are used in mapping to represent the change in magnitude of a variable across geographic space. In some cases the colors start soft and end bold, in others they start on one side of the color wheel and end on the other. In yet others they represent one portion of the color wheel exclusively, which is especially nice for color ramped point features.

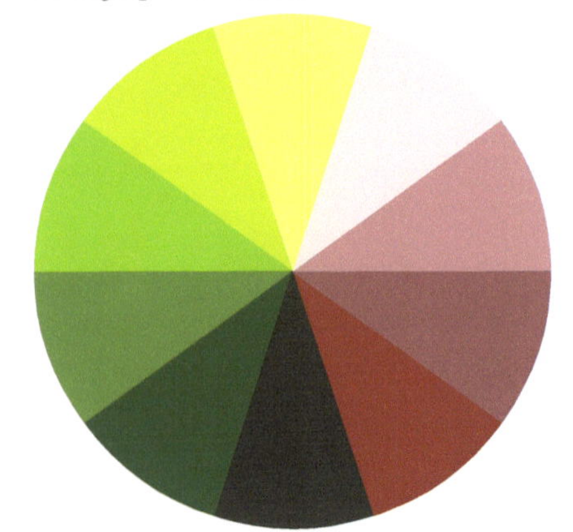

		HEX	RGB	CMYK %
1		003000	0 48 0	79 52 87 68
2		1F4F14	31 79 20	81 42 100 44
3		648744	100 135 68	65 29 92 11
4		94C11C	148 193 28	48 04 100 00
5		C1F203	193 242 3	29 00 100 00
6		F1FF9F	241 255 159	08 00 46 00
7		F9E4E3	249 228 227	01 11 06 00
8		CA9196	202 145 150	21 48 31 00
9		996561	153 101 97	35 63 55 13
10		8E2612	142 38 18	28 94 100 29

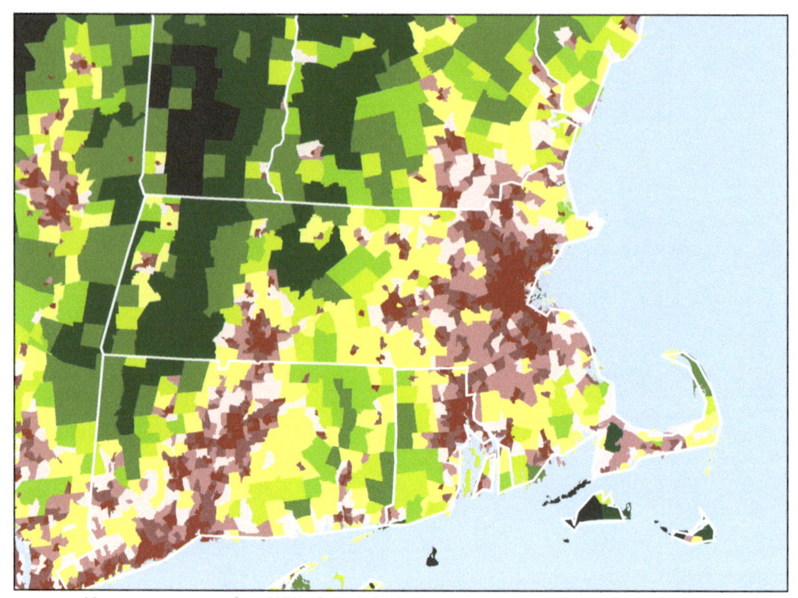

Non-ramp ocean color: HEX C2D1EC RGB 194 209 236 CMYK 22 12 00 00

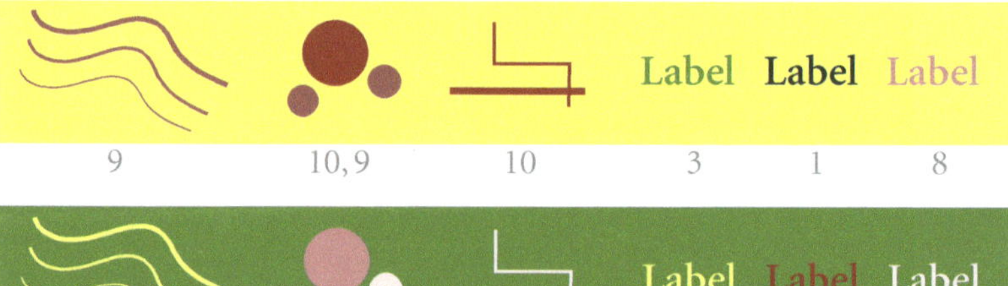

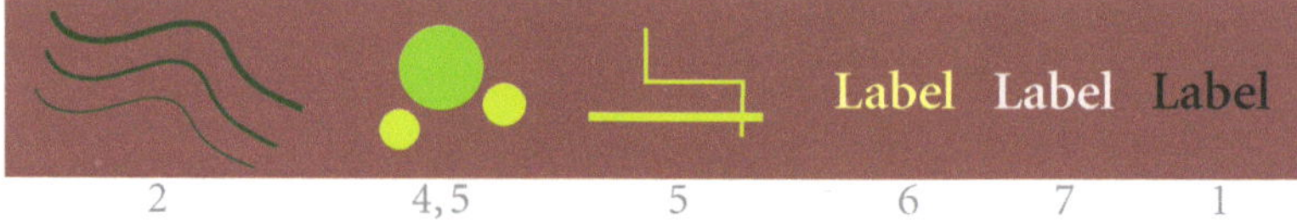

Deuteranope Simulation

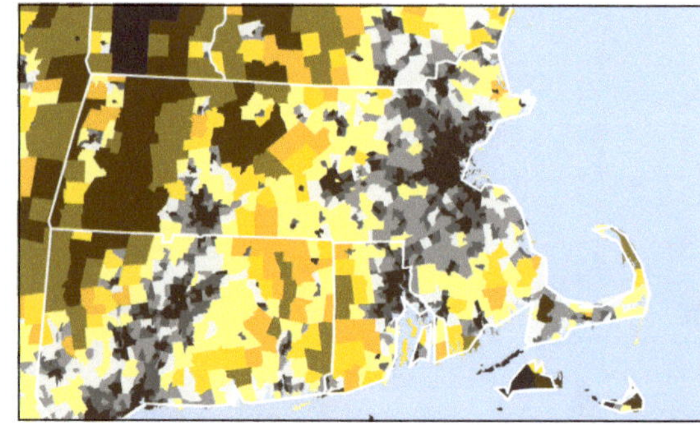

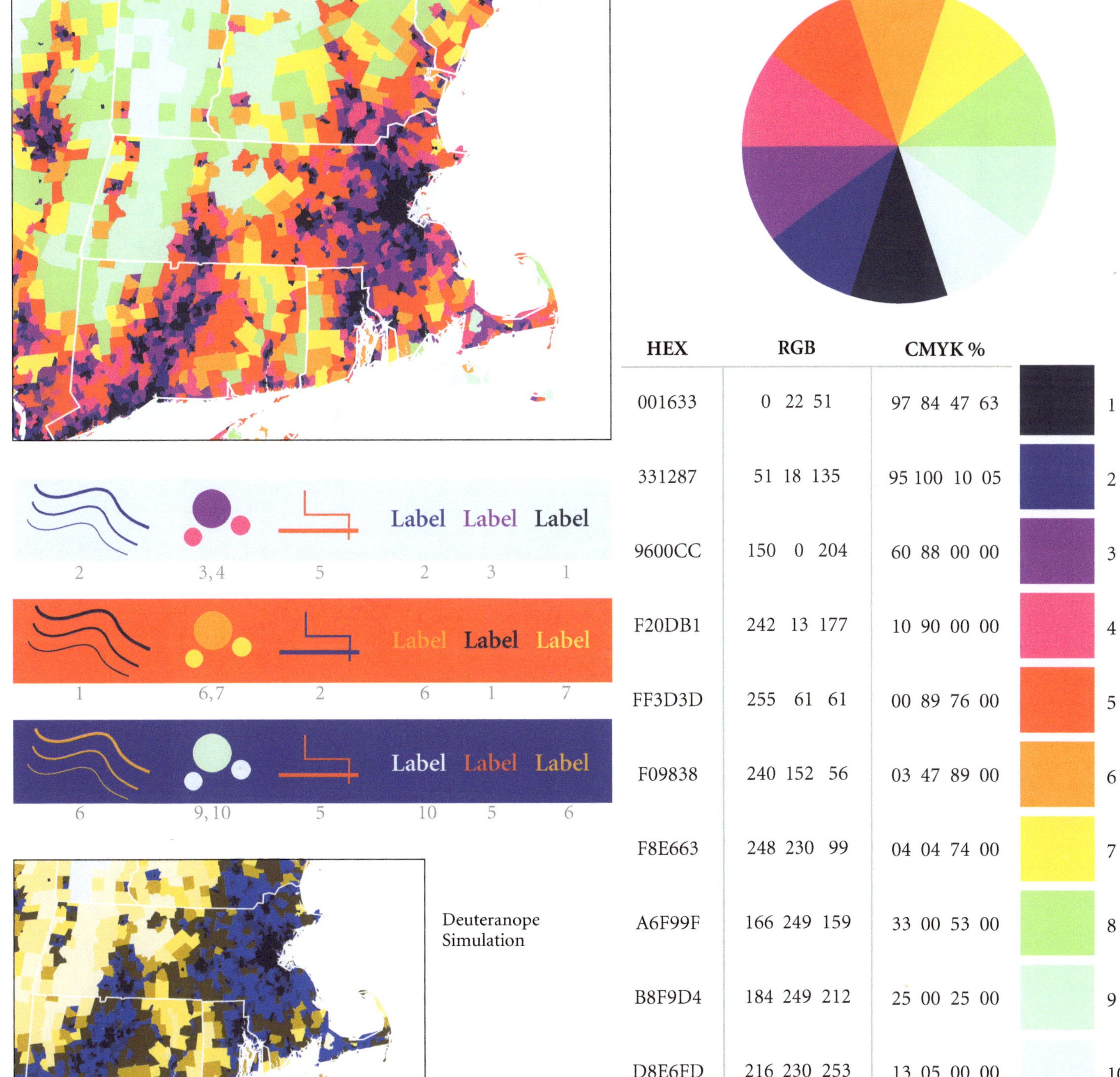

HEX	RGB	CMYK %	
001633	0 22 51	97 84 47 63	1
331287	51 18 135	95 100 10 05	2
9600CC	150 0 204	60 88 00 00	3
F20DB1	242 13 177	10 90 00 00	4
FF3D3D	255 61 61	00 89 76 00	5
F09838	240 152 56	03 47 89 00	6
F8E663	248 230 99	04 04 74 00	7
A6F99F	166 249 159	33 00 53 00	8
B8F9D4	184 249 212	25 00 25 00	9
D8E6FD	216 230 253	13 05 00 00	10

Deuteranope Simulation

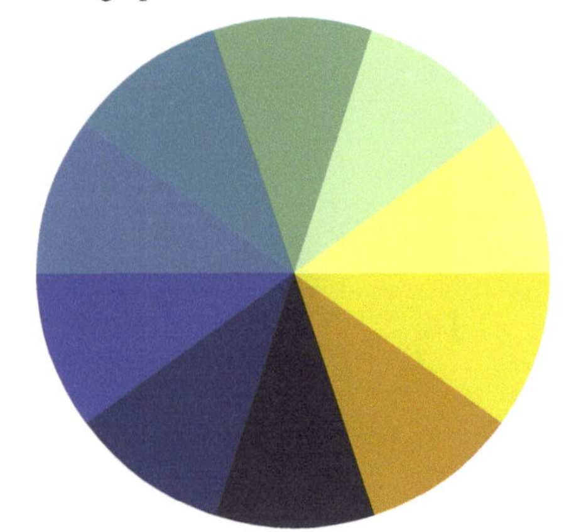

	HEX	RGB	CMYK %
1	1D2B35	29 43 53	84 69 55 60
2	252C5F	37 44 95	99 93 33 25
3	3F4686	63 70 134	89 83 17 04
4	597093	89 112 147	71 53 25 04
5	577C8F	87 124 143	70 43 34 05
6	75A07D	117 160 125	58 22 60 02
7	BCDBAD	188 219 173	27 01 40 00
8	EFFDA3	239 253 163	08 00 44 00
9	DED643	222 214 67	15 07 88 00
10	BD8A37	189 138 55	25 45 94 05

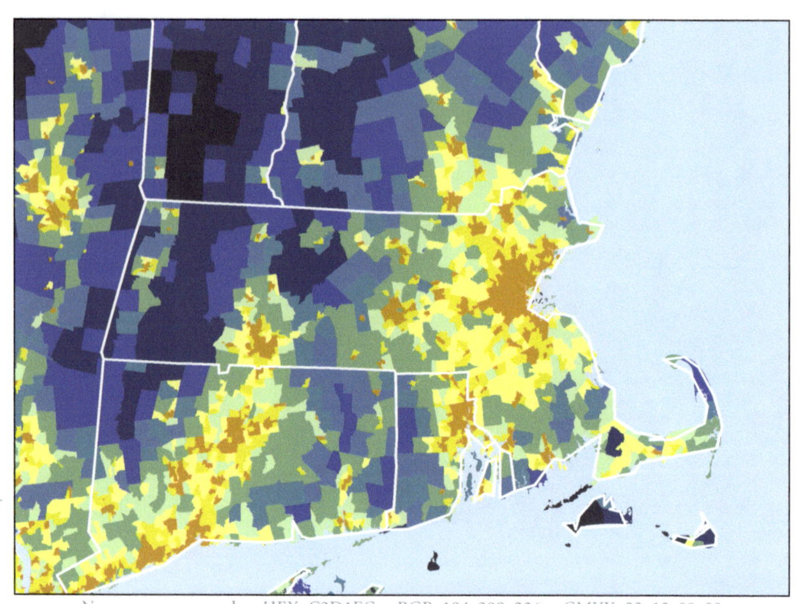

Non-ramp ocean color: HEX C2D1EC RGB 194 209 236 CMYK 22 12 00 00

Label Label Label

6 1, 2 10 10 8 7

Label Label Label

8 10, 9 1 8 3 2

Label Label Label

1 4, 5 10 6 3 2

Deuteranope Simulation

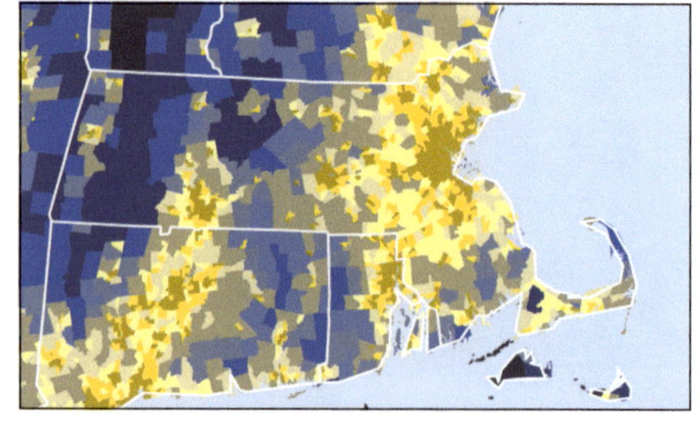

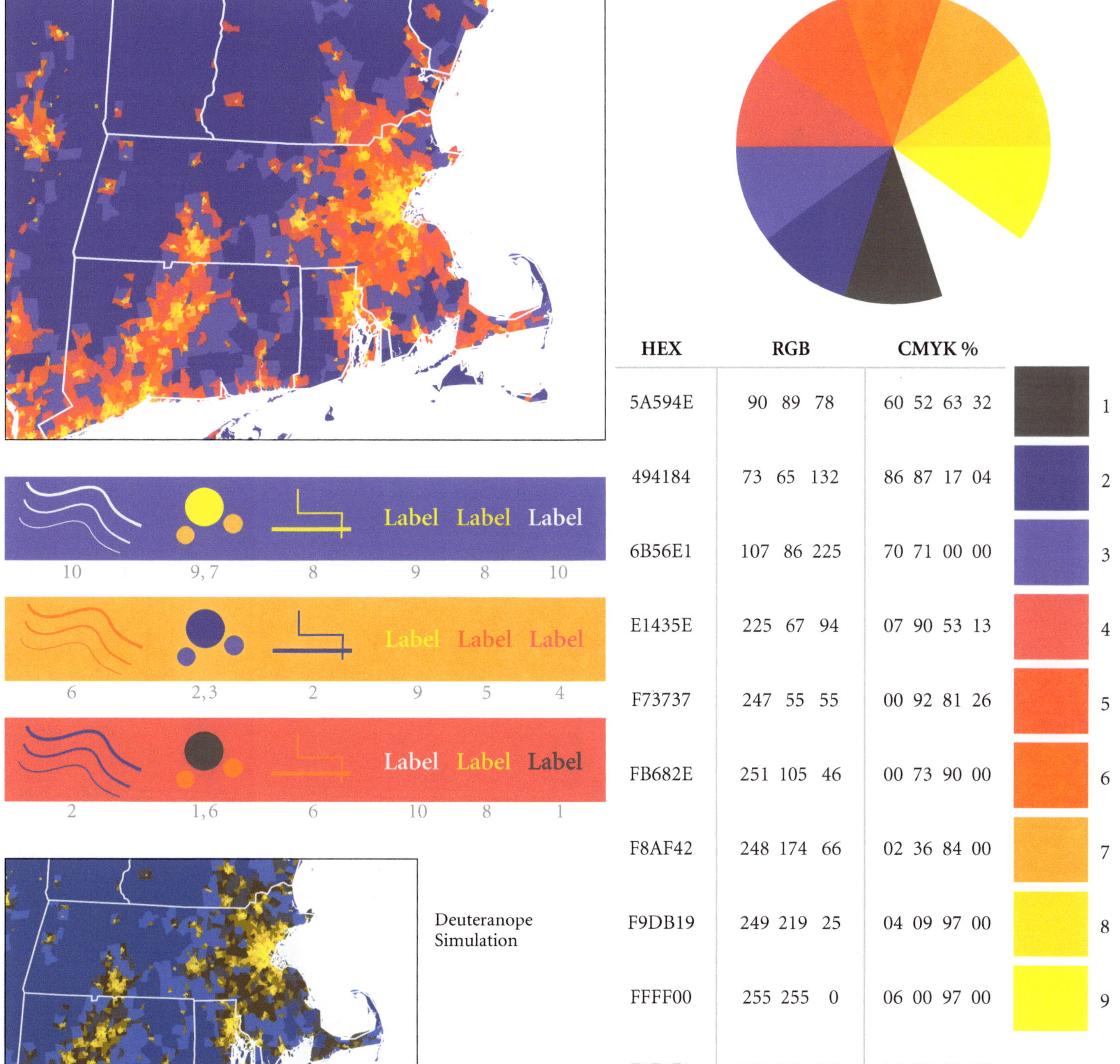

HEX	RGB	CMYK %	
5A594E	90 89 78	60 52 63 32	1
494184	73 65 132	86 87 17 04	2
6B56E1	107 86 225	70 71 00 00	3
E1435E	225 67 94	07 90 53 13	4
F73737	247 55 55	00 92 81 26	5
FB682E	251 105 46	00 73 90 00	6
F8AF42	248 174 66	02 36 84 00	7
F9DB19	249 219 25	04 09 97 00	8
FFFF00	255 255 0	06 00 97 00	9
F2F2F2	242 242 242	04 03 03 00	10

Deuteranope Simulation

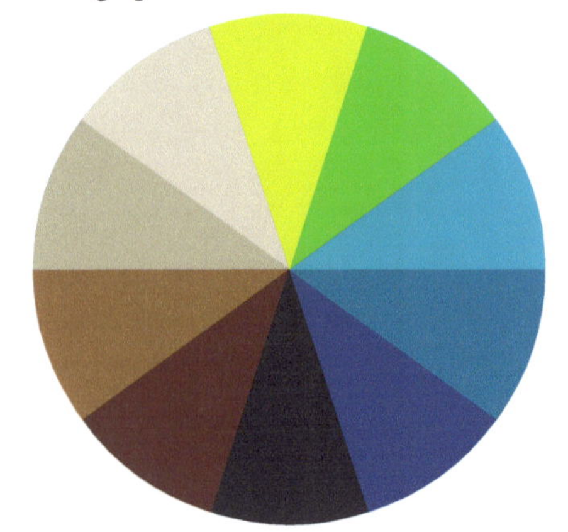

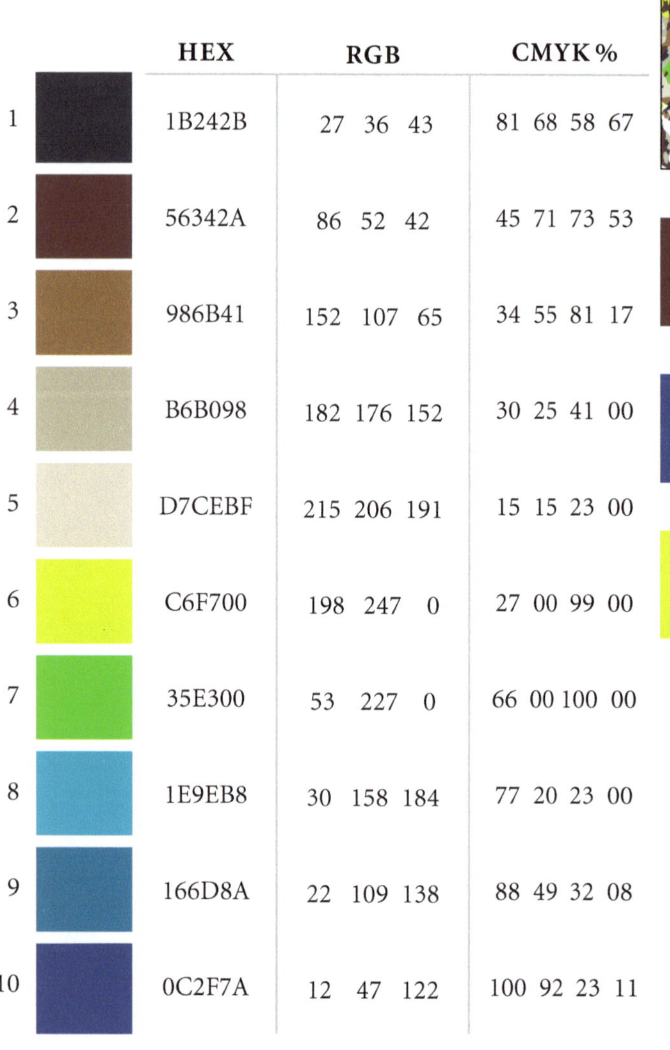

	HEX	RGB	CMYK %
1	1B242B	27 36 43	81 68 58 67
2	56342A	86 52 42	45 71 73 53
3	986B41	152 107 65	34 55 81 17
4	B6B098	182 176 152	30 25 41 00
5	D7CEBF	215 206 191	15 15 23 00
6	C6F700	198 247 0	27 00 99 00
7	35E300	53 227 0	66 00 100 00
8	1E9EB8	30 158 184	77 20 23 00
9	166D8A	22 109 138	88 49 32 08
10	0C2F7A	12 47 122	100 92 23 11

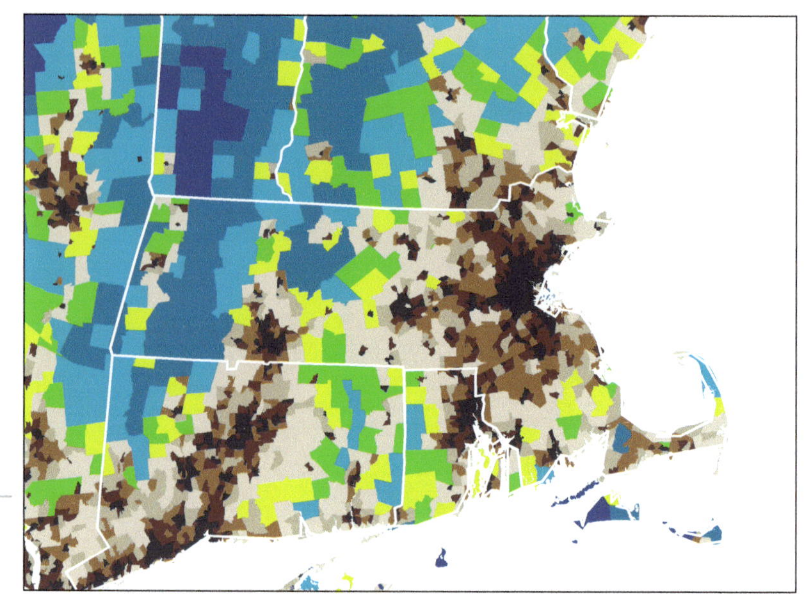

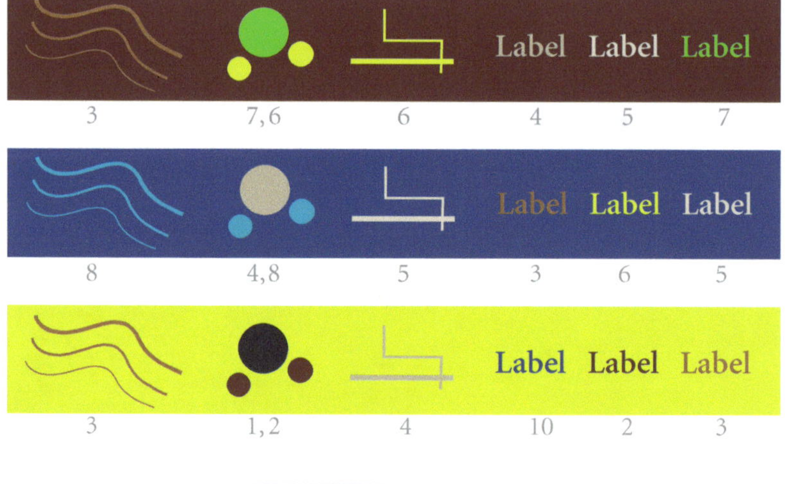

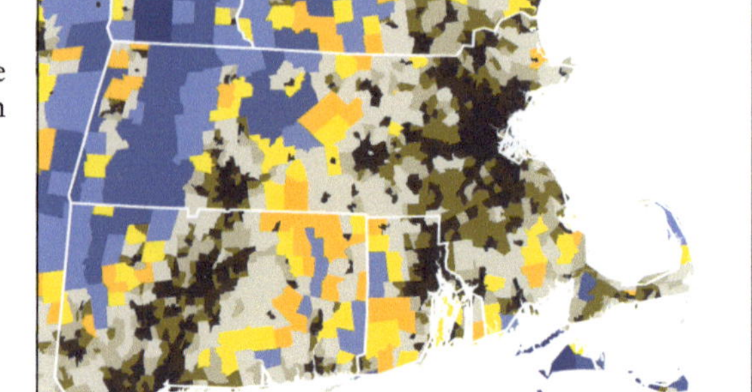

Deuteranope Simulation

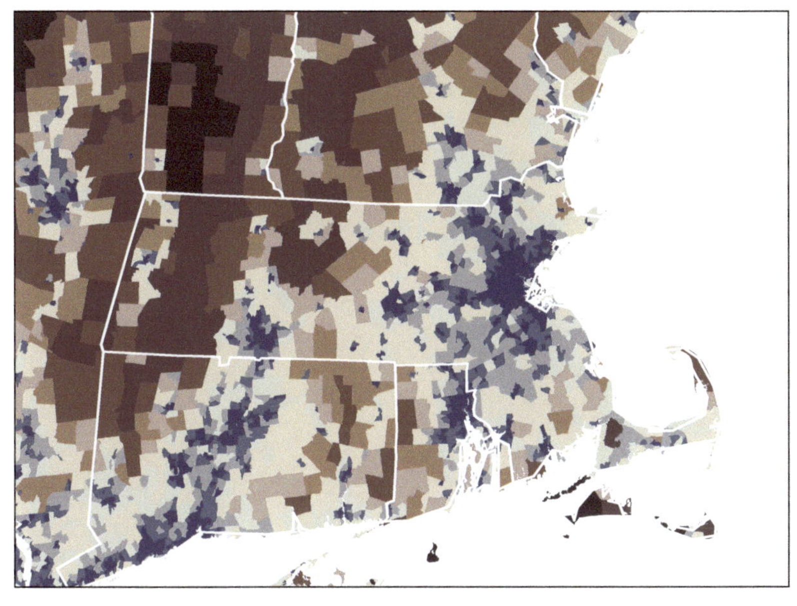

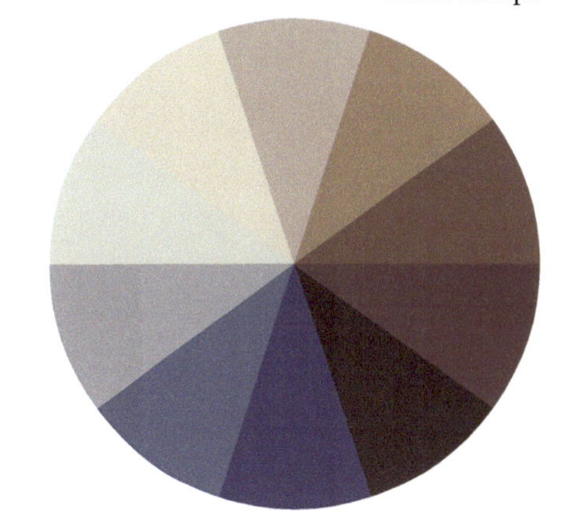

Label Label Label

10 3, 4 5 5 4 7

Label Label Label

8 1, 2 8 1 8 2

Label Label Label

6 10, 9 7 10 1 8

Deuteranope Simulation

HEX	RGB	CMYK %	
403958	64 57 88	79 79 40 32	1
5F6074	95 96 116	66 59 38 15	2
9E9EA6	158 158 166	41 33 28 00	3
CED0C5	206 208 197	19 13 21 00	4
D7CEBF	215 206 191	15 15 23 00	5
BAA496	186 164 150	28 33 39 00	6
A07C62	160 124 98	35 49 63 10	7
755548	117 85 72	44 61 66 30	8
5A463F	90 70 63	53 63 65 42	9
271511	39 21 17	60 72 71 79	10

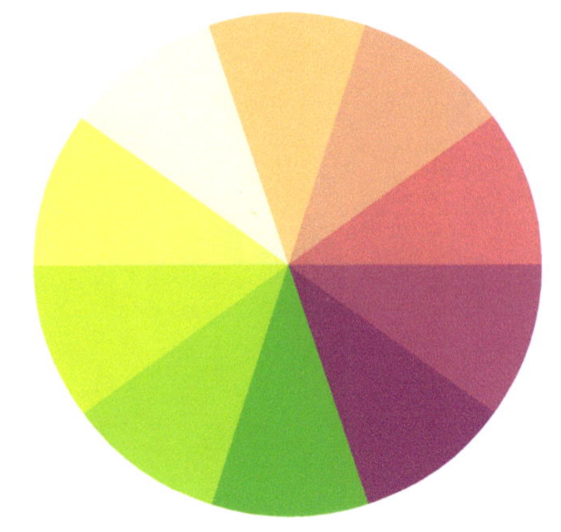

	HEX	RGB	CMYK %
1	59A80F	89 168 15	70 10 100 01
2	9ED54C	158 213 76	42 00 90 00
3	C4ED68	196 237 104	26 00 74 00
4	E2FF9E	226 255 158	13 00 49 00
5	F0F2DD	240 242 221	05 01 14 00
6	F8CA8C	248 202 140	02 22 50 00
7	E9A189	233 161 137	06 42 43 00
8	D47384	212 115 132	14 67 35 01
9	AC437B	172 67 123	33 88 25 02
10	8C286E	140 40 110	48 98 27 08

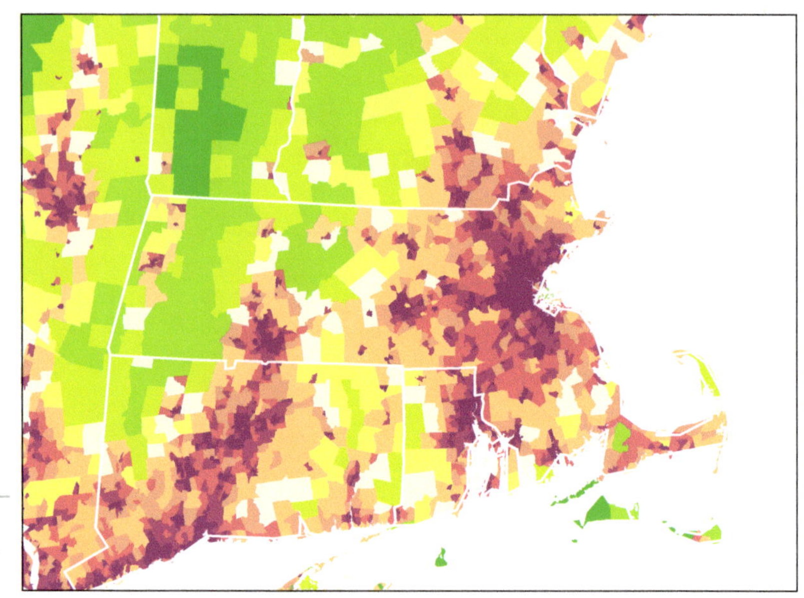

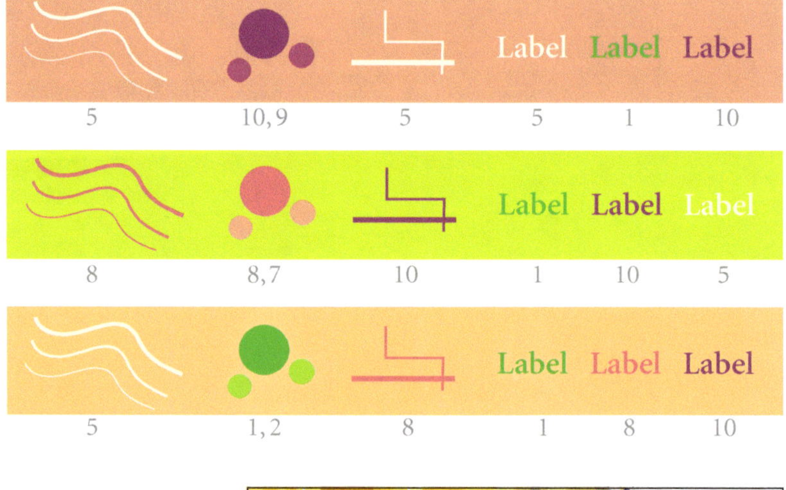

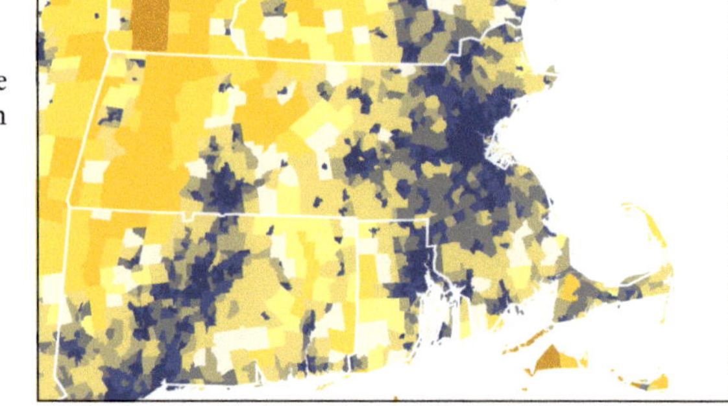

Deuteranope Simulation

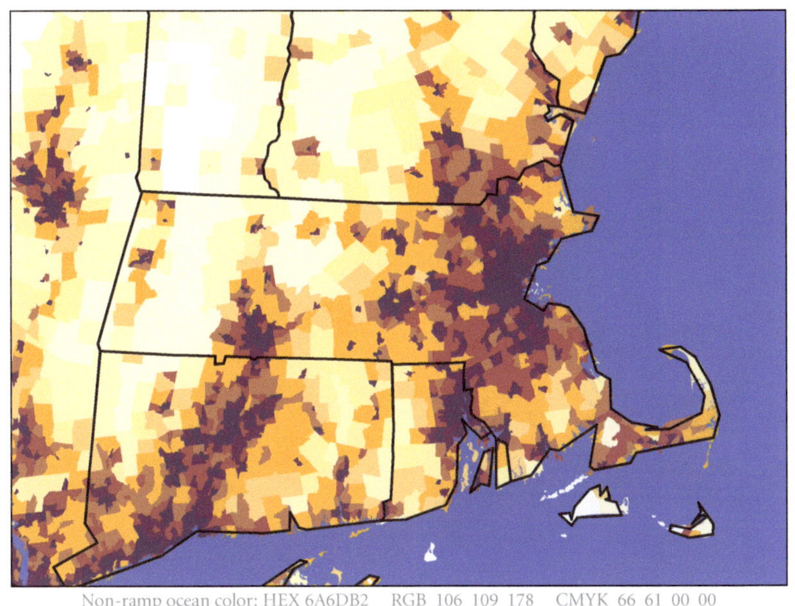

Non-ramp ocean color: HEX 6A6DB2 RGB 106 109 178 CMYK 66 61 00 00

Label Label Label

6 1, 2 3 2 5 1

Label Label Label

10 2, 3 5 9 1 7

Label Label Label

4 3, 4 1 3 2 1

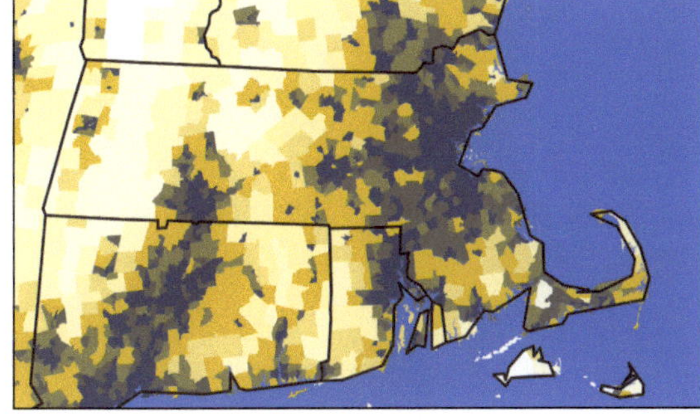

Deuteranope Simulation

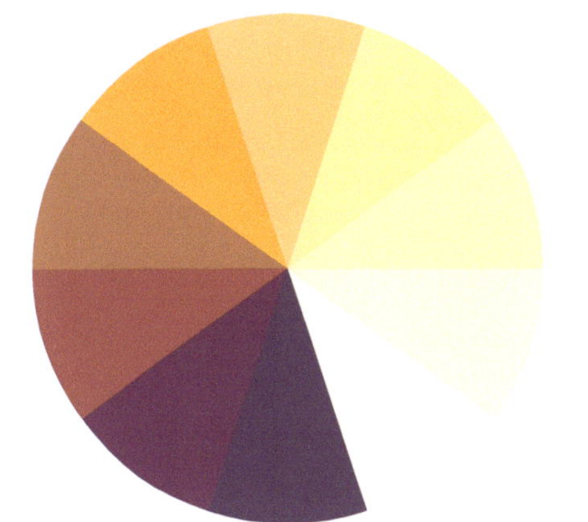

HEX	RGB	CMYK %	
604860	96 72 96	63 73 41 25	1
784860	120 72 96	50 77 42 21	2
A86060	168 96 96	29 69 55 10	3
C07860	192 120 96	22 59 63 04	4
F0A848	240 168 72	04 38 82 00	5
F8CA8C	248 202 140	02 22 50 00	6
FEECAE	254 236 174	01 05 38 00	7
FFF4C2	255 244 194	01 02 28 00	8
FFF7DB	255 247 219	00 02 15 00	9
FFFCF6	255 252 246	00 01 02 00	10

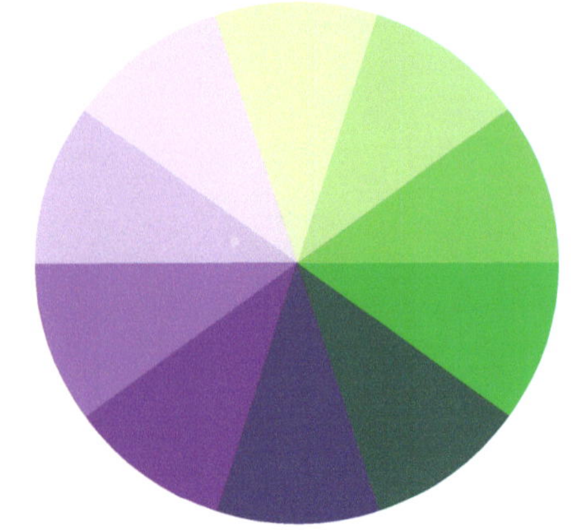

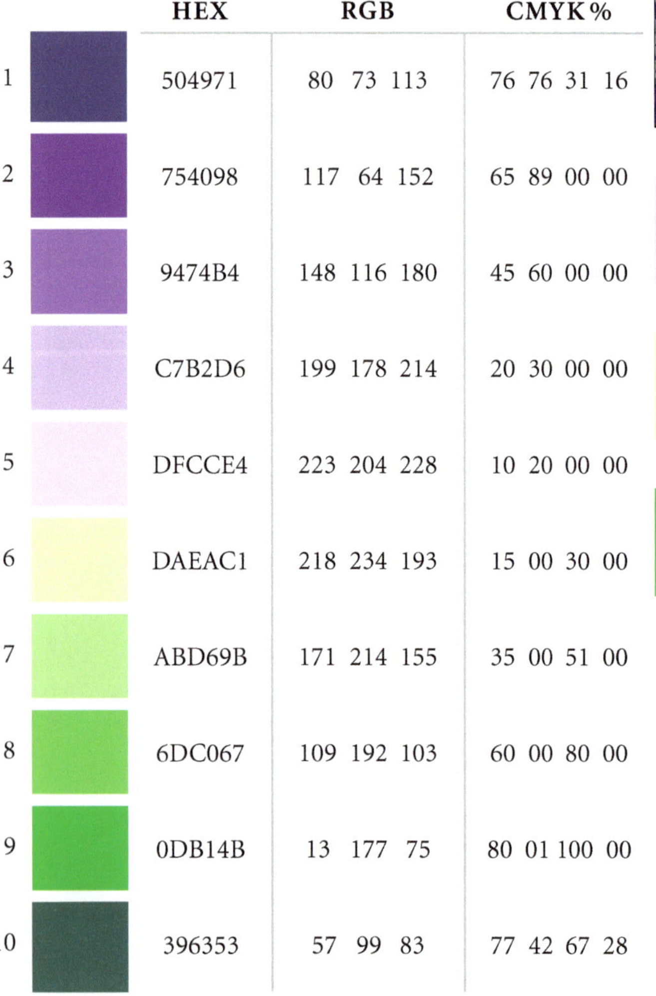

	HEX	RGB	CMYK %
1	504971	80 73 113	76 76 31 16
2	754098	117 64 152	65 89 00 00
3	9474B4	148 116 180	45 60 00 00
4	C7B2D6	199 178 214	20 30 00 00
5	DFCCE4	223 204 228	10 20 00 00
6	DAEAC1	218 234 193	15 00 30 00
7	ABD69B	171 214 155	35 00 51 00
8	6DC067	109 192 103	60 00 80 00
9	0DB14B	13 177 75	80 01 100 00
10	396353	57 99 83	77 42 67 28

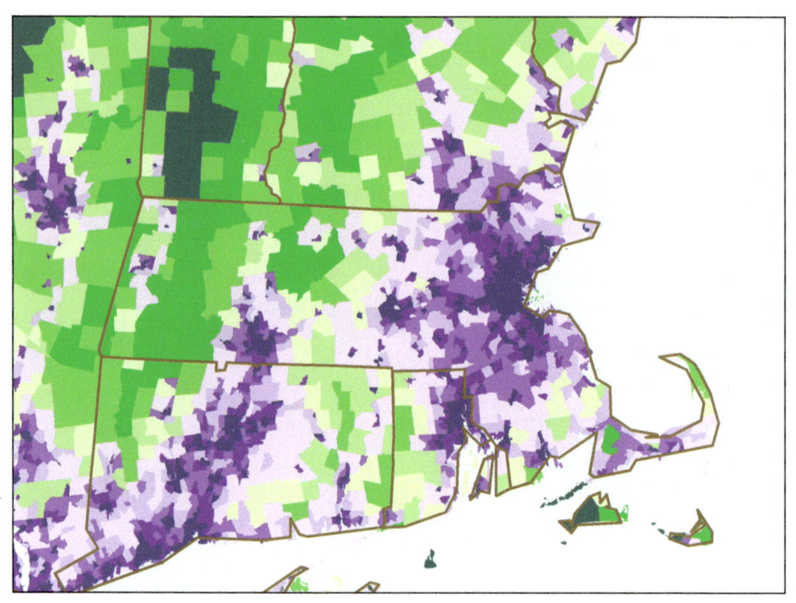

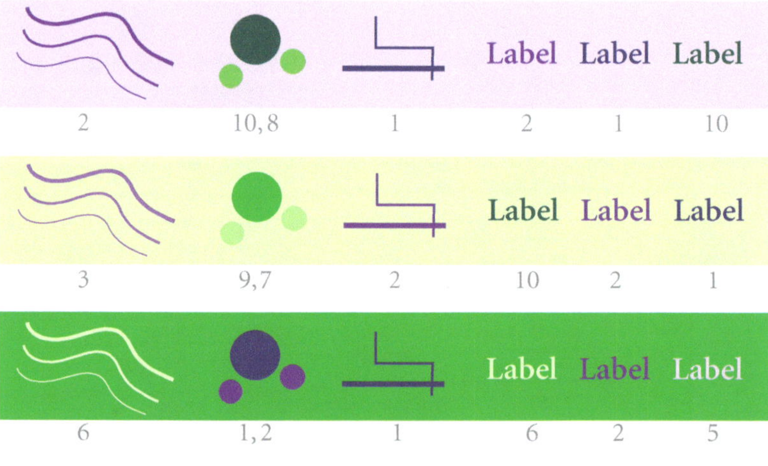

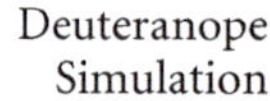

Deuteranope Simulation

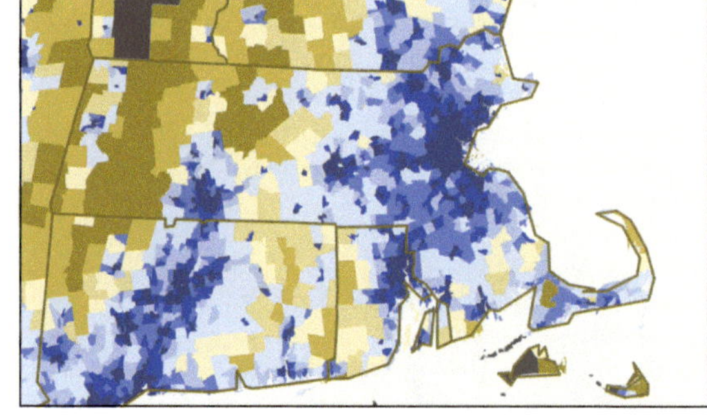

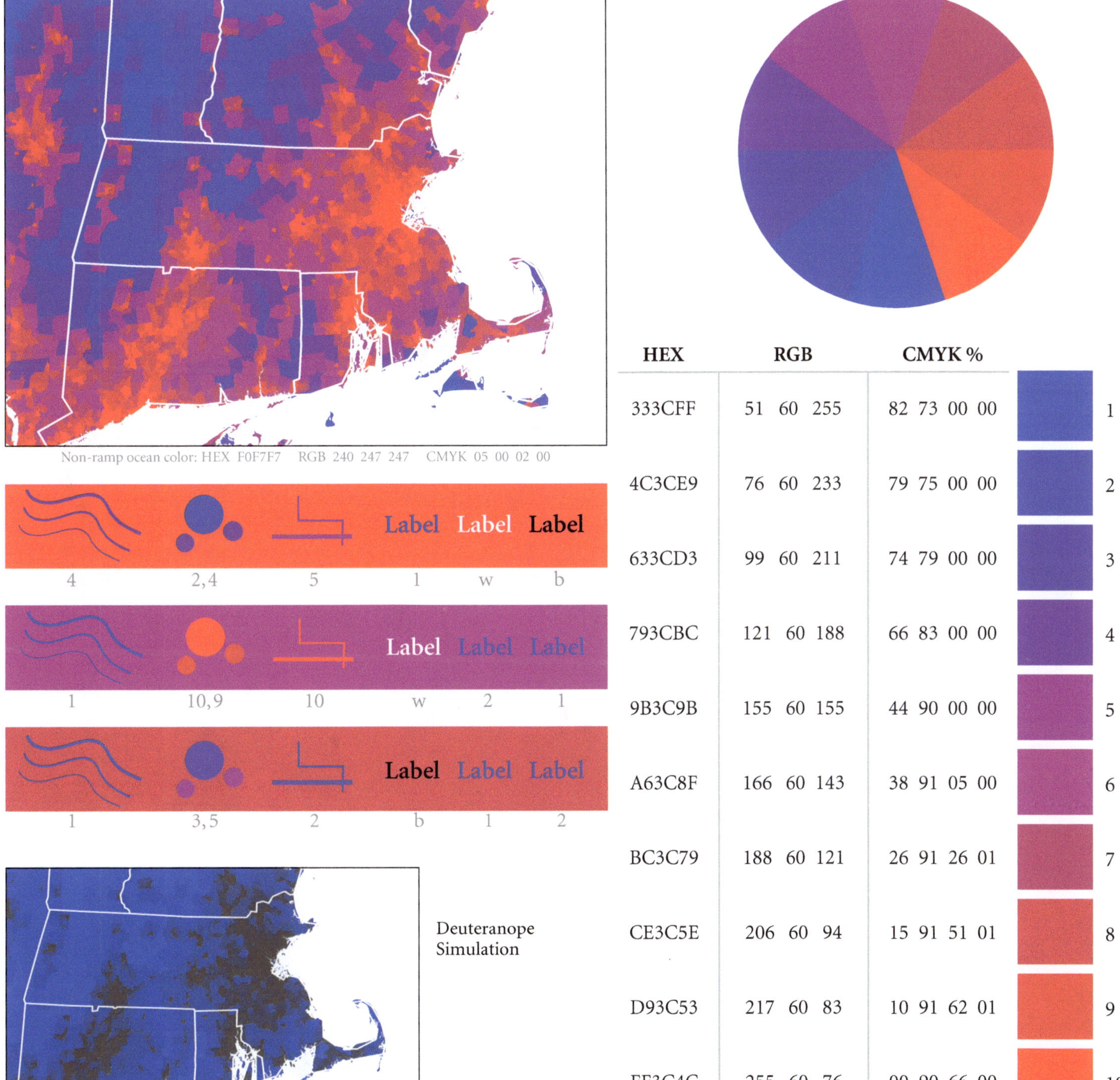

Non-ramp ocean color: HEX F0F7F7 RGB 240 247 247 CMYK 05 00 02 00

Deuteranope Simulation

HEX	RGB	CMYK %	
333CFF	51 60 255	82 73 00 00	1
4C3CE9	76 60 233	79 75 00 00	2
633CD3	99 60 211	74 79 00 00	3
793CBC	121 60 188	66 83 00 00	4
9B3C9B	155 60 155	44 90 00 00	5
A63C8F	166 60 143	38 91 05 00	6
BC3C79	188 60 121	26 91 26 01	7
CE3C5E	206 60 94	15 91 51 01	8
D93C53	217 60 83	10 91 62 01	9
FF3C4C	255 60 76	00 90 66 00	10

TYPOGRAPHY

Typography

Perfect typography is certainly the most elusive of all arts. Sculpture in stone alone comes near it in obstinacy.
—Jan Tschichold

Typography is one of the major aspects of map design that must be undertaken with care and enthusiasm. Along with color, content, message, and layout, it makes a significant impact on the usability of the final product. While it may seem easiest to simply stick with Arial or to use the default font in your cartography software, the level of professionalism of the map may suffer. Cartographic font choices must be made deliberately, with a strong primary emphasis placed on readability while also staying mindful of personality and quality.

With myriad font choices available, however, a map maker does not always have an easy means to compare and contrast them for readability, personality, and quality, which then leads the map maker to save time by using a default font. This chapter aims to change that by showcasing 50 high-quality fonts in a way in which they can be quickly, yet thoroughly, compared.

Many of the typeface families shown in this chapter are either free or come standard on a typical PC operating system. Not all systems will have all the standard fonts but it is hoped that enough are shown of the ones that could be available that something appropriate can be chosen. There are also a great many high-quality fonts available for free or for purchase. A selection of these, chosen for their usefulness on maps in particular, are also showcased in this chapter.

To find out if you have one of the fonts showcased in this booklet, open the software you are using to create the map and look for it by name in the font picklist. If it is not there, search for the name of the font on the web to locate a download page. If the font will be for commercial use, check the end user licensing agreement, if any, to determine if this is allowed. Often, there are several sites from which you can download or purchase a font.

Important Definitions

TYPEFACE and **FONT** These terms do have distinct meanings though they are often used interchangeably in both today's colloquial dialog and in this booklet. Traditionally, typeface has referred to a group of letterforms (their aesthetic) while font has referred to the press or computer code that produces them.

SERIF A serif font has small finishing strokes protruding from the edges of most letterforms while a sans serif font does not. We often think of serif fonts as being easier to read in book print—in large blocks of text—because the serifs help to move the eye along the line of text. On computer screens, however, sans serif fonts used to be standard because they are readable at lower resolutions. Sans serif fonts, which some believe to be more legible at all sizes, are traditionally viewed as being good for small amounts of text such as attribute labels and titles on printed media. However, we are seeing these two being combined and swapped in all kinds of media today due to higher-resolution viewing devices and changing aesthetic preferences.

LEADING Leading (pronounced with a short e sound) is the spacing between the lines of text. In maps, this is often tweaked to move a multi-line label closer together or further apart depending on the associated feature's size. It is also referred to as line spacing or interline spacing.

TRACKING Tracking is the spacing between letters. It is also called character spacing in some mapping software. This can be increased to provide emphasis for map text such as titles and to increase readability for light-colored text on dark backgrounds. It can be decreased to better fit a label onto a feature but it is wise to use a condensed version of the font for this instead, if one is available.

ITALIC A typeface with a true italic contains very individual letterforms that are different from the regular version of the typeface. A sloped Roman (also called oblique) is a simulated italic that differs from the regular typeface only in the slope of the letters. The italic function in most mapping and graphics software will use the true italic version of the font if one is available or a simulation if one is not available. Some software will not produce a simulated italic, since the results can be less than ideal.

Map Typography

The most popular typefaces are the easiest to read; their popularity has made them disappear from conscious cognition. It becomes impossible to tell if they are easy to read because they are commonly used, or if they are commonly used because they are easy to read. –Zuzana Licko

The cartographic implementation of typography is somewhat different from the traditional typography needs for books and newsprint. In mapping, text is generally one or two words long and under extremes of space constraints. Add to this that the typical map reader is not necessarily familiar with most or all of the text on a map, reducing the ability to discern words through context, and we have some special considerations that must be made when thinking about what typefaces will work on maps.

READABILITY AT SMALL SIZES Most maps need feature labels, and in many cases these labels need to be displayed at very small sizes. While sizes less than 6 pts are generally not legible no matter what typeface is used, attribute labels at 6 or 7 pts are common. At those text sizes it is extremely important that the typeface chosen has the following: tall x-heights (i.e., the lower case letters are taller than usual); open counters (e.g., the o and the e will not fill in with ink); and no serifs.

COMBINING TYPEFACES While some maps contain only variants of one sans serif face, it is not uncommon to combine a sans serif typeface with a serif typeface for titling and other text. It is the designer's choice whether to pick two compatible sans serif typefaces—which can be difficult to do—or to choose two compatible faces, with one each of sans and serif. Compatibility is very subjective but may be achieved by choosing typefaces with similar widths, though not necessarily similar weights. A high level of contrast to emphasize the difference between, for example, man made features, usually labeled in sans serif, and natural features, usually labeled in serif, is advised.

MOOD Keep in mind the mood of the map. Attribute labels on common street maps, for example, should become nearly invisible in the sense that the font should have very little personality. The cartographer often, but not always, wishes to choose a typeface that has the following characteristics: seriousness or credibility, non-obtrusiveness, and perhaps a subtle uniqueness.

How To Use This Chapter

Each double-page spread in this chapter showcases two typefaces, one serif and one sans serif. The serif typeface is shown at the top of the right-hand page and the sans serif is shown at the bottom of the right-hand page. The typeface designer is attributed next to the name of the typeface, followed by a short description, all written in the font itself. Under the description is a table of letter and number forms also displayed in the typeface. To the right of the typeface description is a label hierarchy mock-up using the typeface at varying point sizes that simulates a map label hierarchy. Short nonsense sentences, comprised of every letter in the alphabet, demonstrate up to four variants of the typeface as well. A clip of the map is also shown.

The map on the left-hand page has labels in both typefaces. The serif typeface is used for the map title, the three park labels, and the creek and bay labels. The sans serif is used for the road name labels and the town label. The text sizes, tracking, and line spacing may vary slightly from one map to another since each typeface is slightly different in terms of its ability to fit into the map shapes appropriately. The sizes are: title, 21-25 pt; roads, 7-9 pt; parks, 8-10 pt; creek, 10-11 pt; bay, 11-15 pt; town, 25-30 pt.

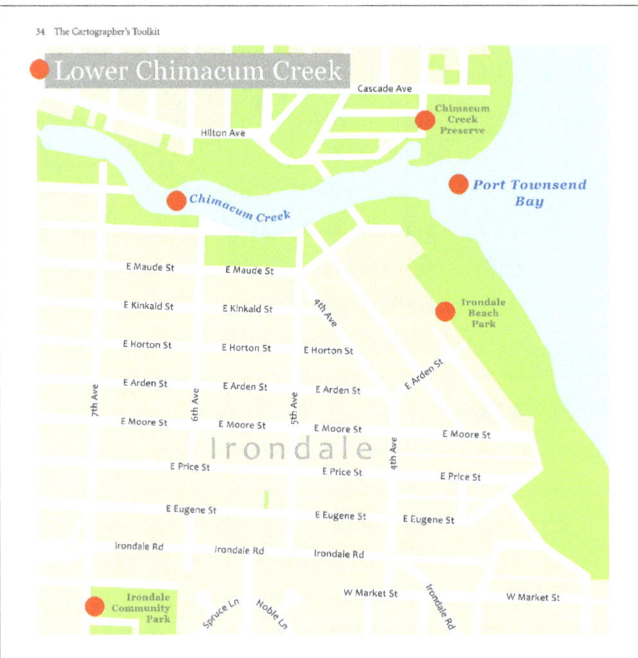

Red dots denote text in the serif font featured on the page. Other map text is in the sans serif font.

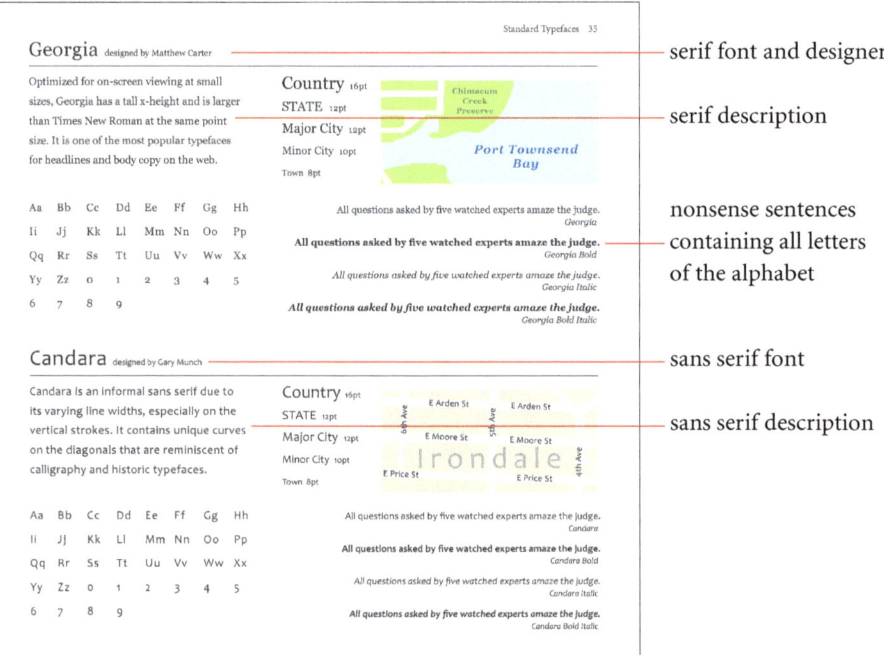

Additional Resources

This chapter is meant to provide a reference for choosing map typefaces and is not meant to be an exhaustive primer on typography technique. For learning more about typography in general, read *Thinking with Type, 2nd revised and expanded edition: A Critical Guide for Designers, Writers, Editors & Students* (Design Briefs) by Ellen Lupton.

Places to buy and download the fonts found here include, but are not limited to: Fontshop, dafont.com, MyFonts, Linotype, exljbris, and fonts.com. Also of interest is the League of Moveable Type.

Type Brewer, by Ben Sheesley, is an online tool for selecting typefaces that is aimed at map makers. The typefaces shown in the program are often Mac OS system fonts or for-fee fonts.

STANDARD TYPEFACES

While no typeface is found across all systems, the ones showcased in this section are typical on PC operating systems. There is a good possibility that these exist on a contemporary GIS machine. They can also be purchased online, if needed.

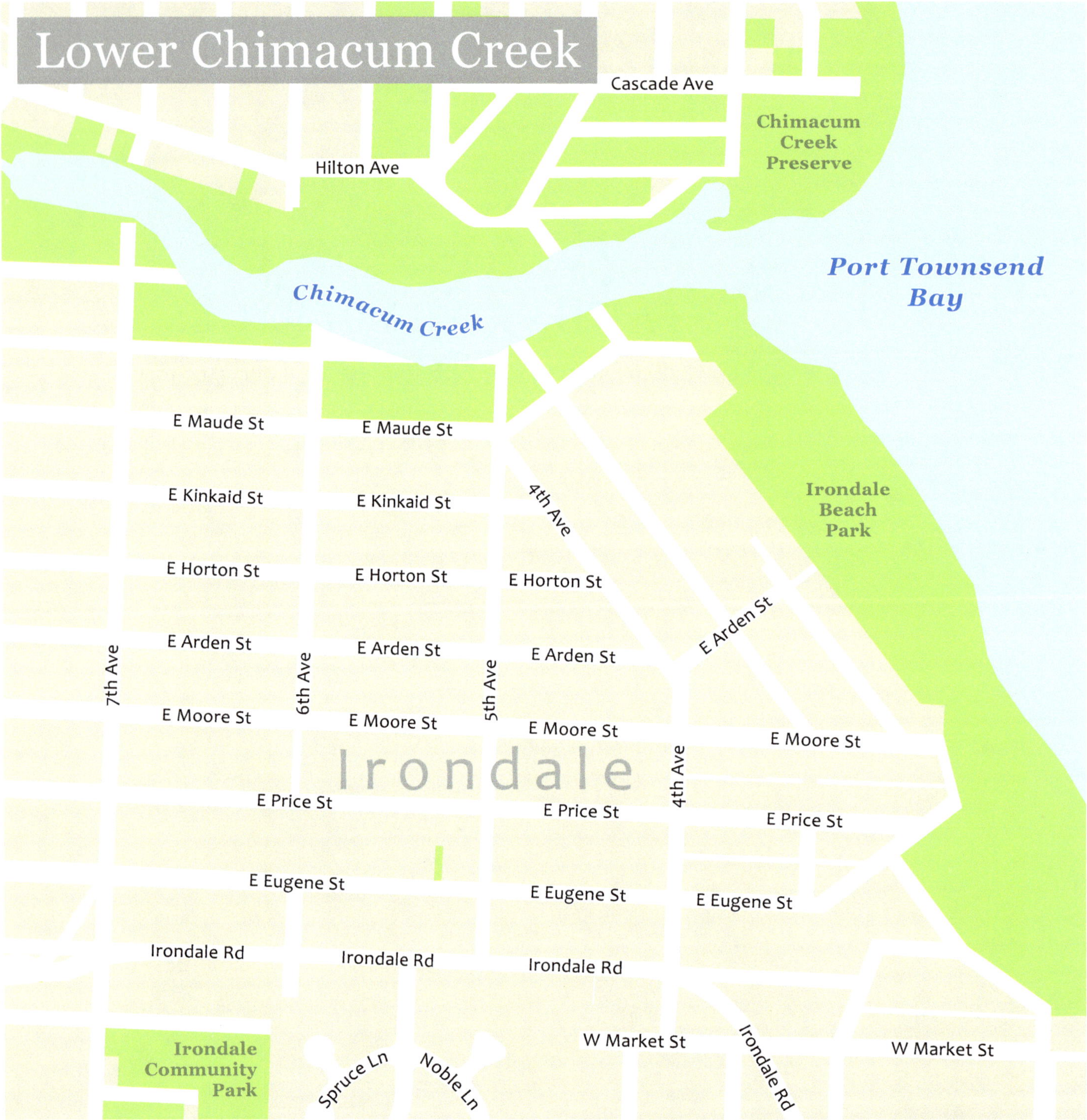

Lower Chimacum Creek

Cascade Ave
Chimacum Creek Preserve
Hilton Ave
Port Townsend Bay
Chimacum Creek
E Maude St
E Maude St
E Kinkaid St
E Kinkaid St
4th Ave
Irondale Beach Park
E Horton St
E Horton St
E Horton St
E Arden St
E Arden St
E Arden St
E Arden St
7th Ave
6th Ave
5th Ave
E Moore St
E Moore St
E Moore St
E Moore St
Irondale
4th Ave
E Price St
E Price St
E Price St
E Eugene St
E Eugene St
E Eugene St
Irondale Rd
Irondale Rd
Irondale Rd
W Market St
W Market St
Irondale Community Park
Spruce Ln
Noble Ln
Irondale Rd

Georgia designed by Matthew Carter

Optimized for on-screen viewing at small sizes, Georgia has a tall x-height and is larger than Times New Roman at the same point size. It is one of the most popular typefaces for headlines and body copy on the web.

Country 16pt
STATE 12pt
Major City 12pt
Minor City 10pt
Town 8pt

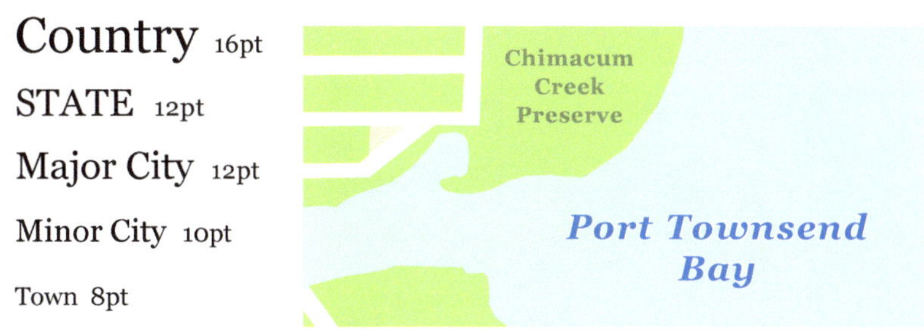

Aa	Bb	Cc	Dd	Ee	Ff	Gg	Hh
Ii	Jj	Kk	Ll	Mm	Nn	Oo	Pp
Qq	Rr	Ss	Tt	Uu	Vv	Ww	Xx
Yy	Zz	0	1	2	3	4	5
6	7	8	9				

All questions asked by five watched experts amaze the judge.
Georgia

All questions asked by five watched experts amaze the judge.
Georgia Bold

All questions asked by five watched experts amaze the judge.
Georgia Italic

All questions asked by five watched experts amaze the judge.
Georgia Bold Italic

Candara designed by Gary Munch

Candara is an informal sans serif due to its varying line widths, especially on the vertical strokes. It contains unique curves on the diagonals that are reminiscent of calligraphy and historic typefaces.

Country 16pt
STATE 12pt
Major City 12pt
Minor City 10pt
Town 8pt

E Arden St
E Arden St
6th Ave
5th Ave
E Moore St
E Moore St
Irondale
4th Ave
E Price St
E Price St

Aa	Bb	Cc	Dd	Ee	Ff	Gg	Hh
Ii	Jj	Kk	Ll	Mm	Nn	Oo	Pp
Qq	Rr	Ss	Tt	Uu	Vv	Ww	Xx
Yy	Zz	0	1	2	3	4	5
6	7	8	9				

All questions asked by five watched experts amaze the judge.
Candara

All questions asked by five watched experts amaze the judge.
Candara Bold

All questions asked by five watched experts amaze the judge.
Candara Italic

All questions asked by five watched experts amaze the judge.
Candara Bold Italic

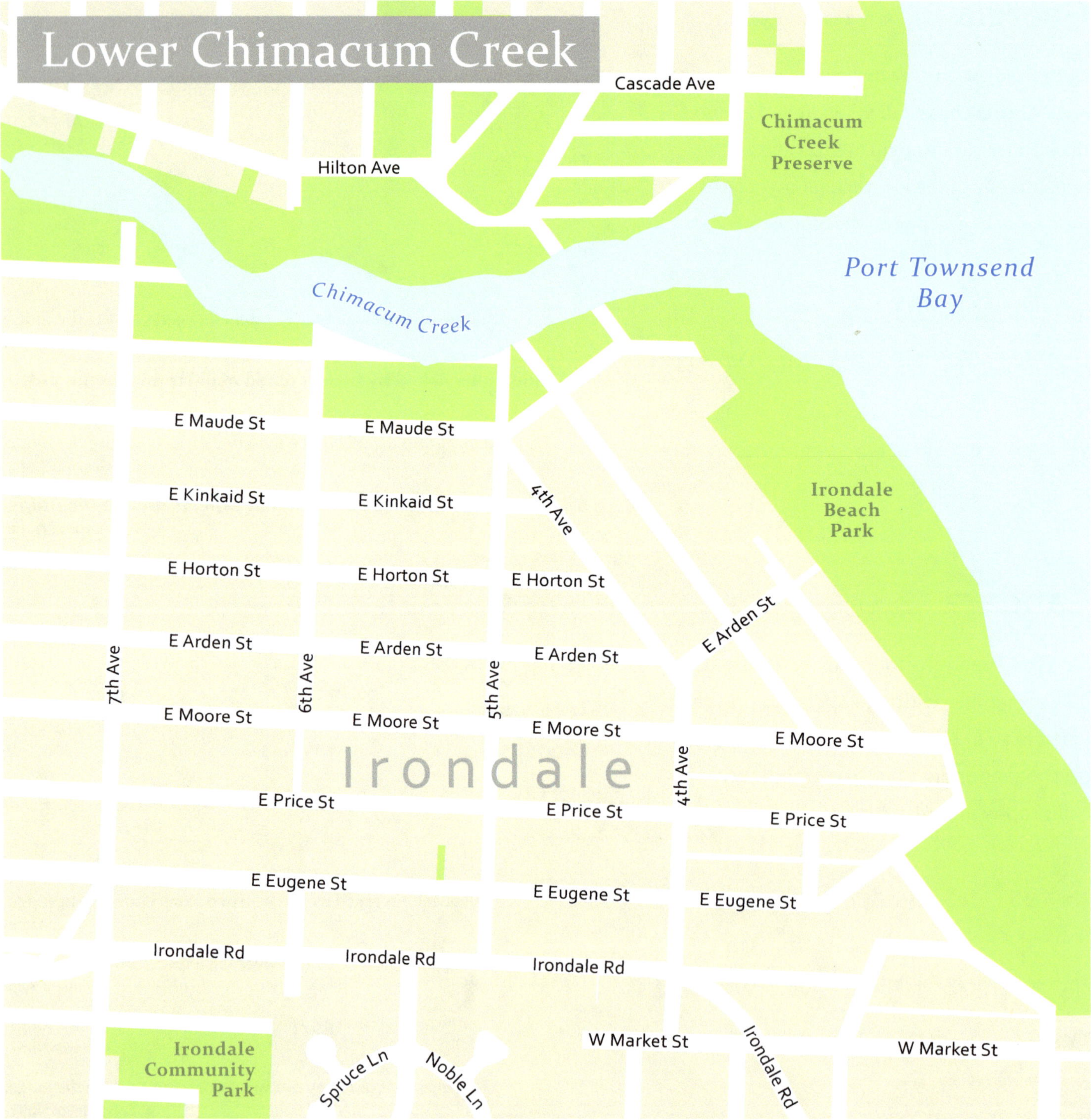

Lower Chimacum Creek

Cascade Ave
Chimacum Creek Preserve
Hilton Ave
Port Townsend Bay
Chimacum Creek
E Maude St
E Maude St
4th Ave
Irondale Beach Park
E Kinkaid St
E Kinkaid St
E Horton St
E Horton St
E Horton St
E Arden St
E Arden St
E Arden St
E Arden St
7th Ave
6th Ave
5th Ave
E Moore St
E Moore St
E Moore St
E Moore St
Irondale
4th Ave
E Price St
E Price St
E Price St
E Eugene St
E Eugene St
E Eugene St
Irondale Rd
Irondale Rd
Irondale Rd
W Market St
W Market St
Irondale Rd
Irondale Community Park
Spruce Ln
Noble Ln

Constantia designed by John Hudson

Optimized for readability in both on-screen viewing and print, Constantia is suitable for maps that have dual functions: to be viewed on the web and printed. Its personality is a compromise between antique and modern.

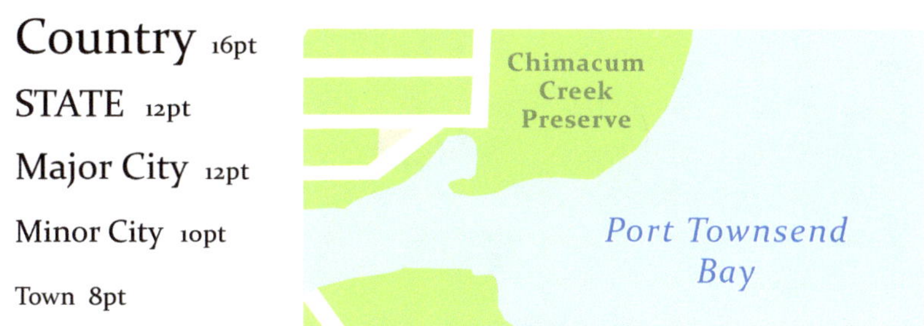

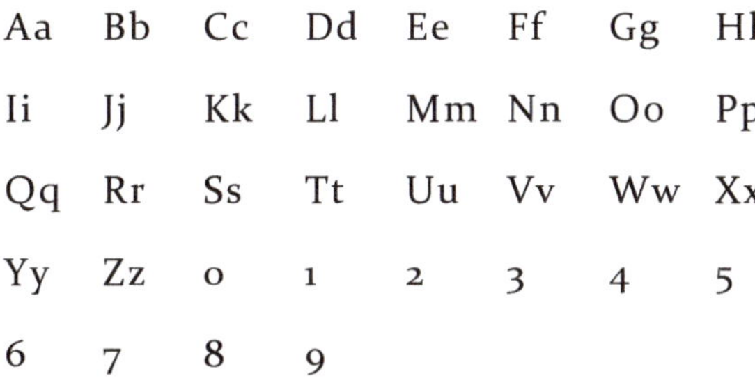

All questions asked by five watched experts amaze the judge.
Constantia

All questions asked by five watched experts amaze the judge.
Constantia Bold

All questions asked by five watched experts amaze the judge.
Constantia Italic

All questions asked by five watched experts amaze the judge.
Constantia Bold Italic

Corbel designed by Jeremy Tankard

Corbel is suitable for on-screen viewing at small sizes. With its clean and consistent strokes and modern geometric shapes, it has an uncluttered look. Its flowing curves soften its modernity.

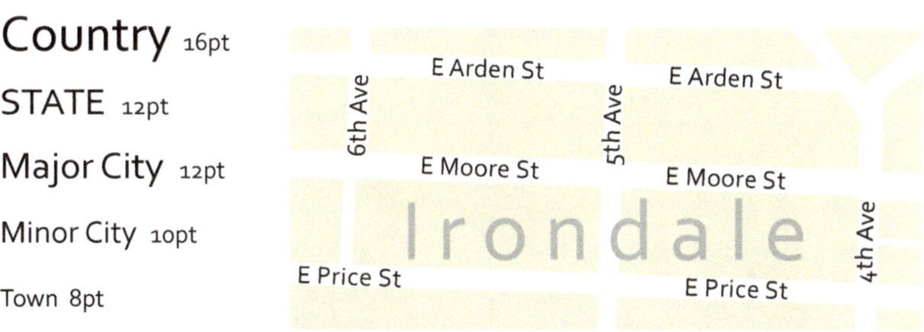

Aa Bb Cc Dd Ee Ff Gg Hh
Ii Jj Kk Ll Mm Nn Oo Pp
Qq Rr Ss Tt Uu Vv Ww Xx
Yy Zz 0 1 2 3 4 5
6 7 8 9

All questions asked by five watched experts amaze the judge.
Corbel

All questions asked by five watched experts amaze the judge.
Corbel Bold

All questions asked by five watched experts amaze the judge.
Corbel Italic

All questions asked by five watched experts amaze the judge.
Corbel Bold Italic

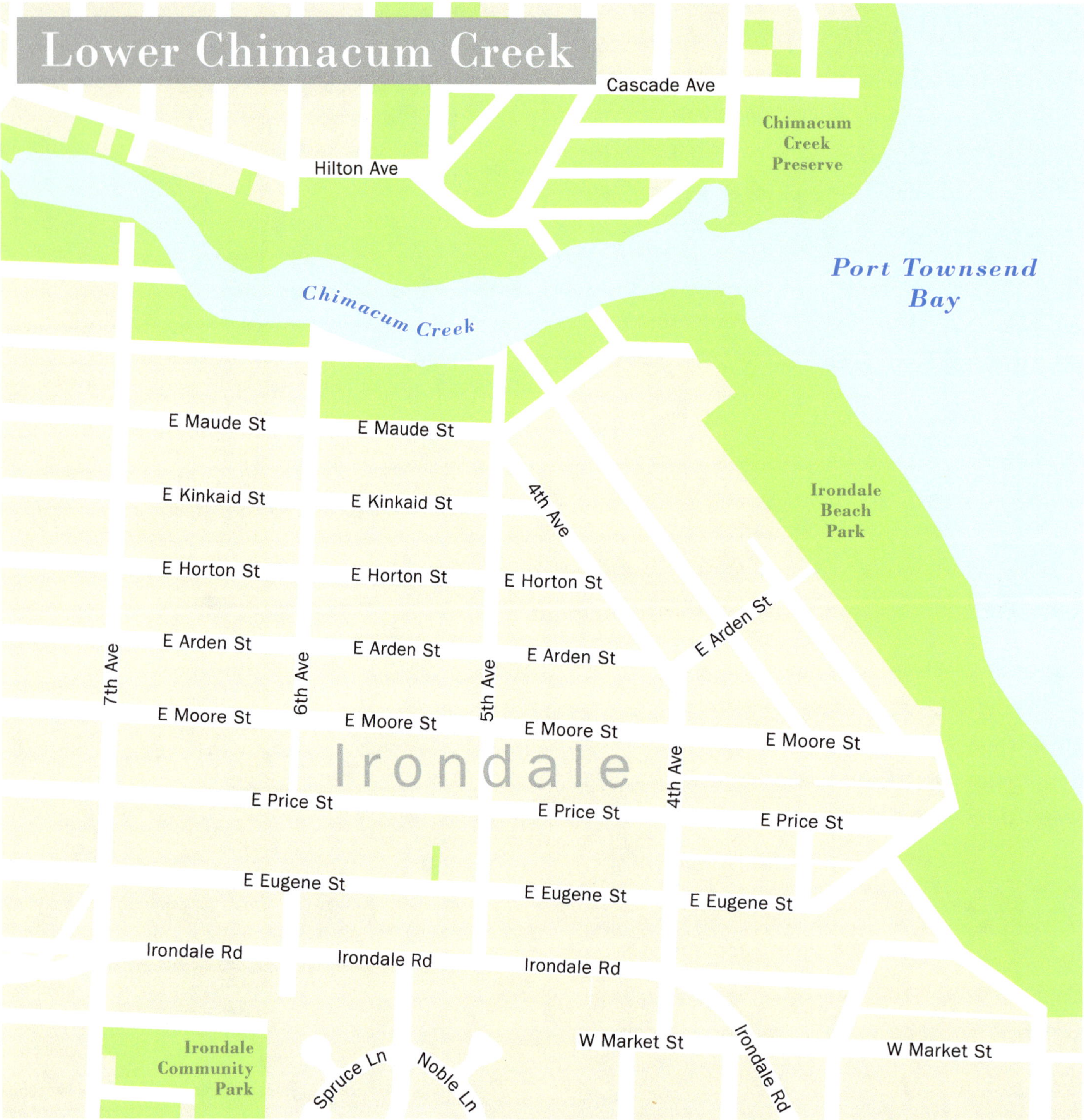

Lower Chimacum Creek

Cascade Ave
Chimacum Creek Preserve
Hilton Ave
Port Townsend Bay
Chimacum Creek
E Maude St
E Maude St
E Kinkaid St
E Kinkaid St
4th Ave
Irondale Beach Park
E Horton St
E Horton St
E Horton St
E Arden St
E Arden St
E Arden St
E Arden St
E Arden St
7th Ave
6th Ave
5th Ave
E Moore St
E Moore St
E Moore St
E Moore St
Irondale
4th Ave
E Price St
E Price St
E Price St
E Eugene St
E Eugene St
E Eugene St
Irondale Rd
Irondale Rd
Irondale Rd
W Market St
W Market St
Irondale Rd
Irondale Community Park
Spruce Ln
Noble Ln

Bodoni MT designed by Giambattista Bodoni

Bodoni MT's exaggerated verticals make the uppercase letters appear to be condensed. It works best at larger point sizes but does have condensed and poster condensed versions. Its long descenders make it suboptimal for stream labeling.

Country 16pt
STATE 12pt
Major City 12pt
Minor City 10pt
Town 8pt

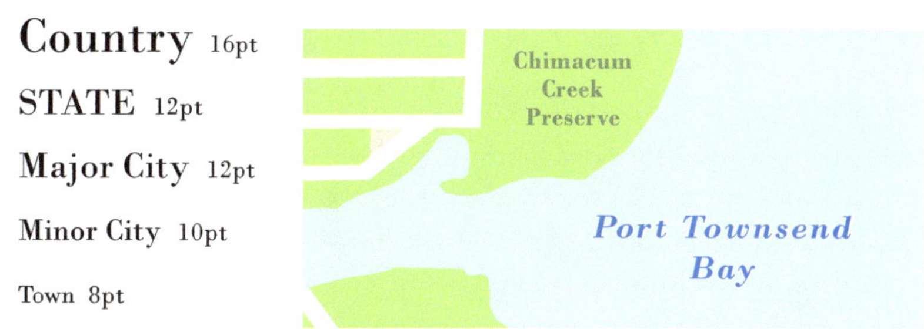

Aa Bb Cc Dd Ee Ff Gg Hh
Ii Jj Kk Ll Mm Nn Oo Pp
Qq Rr Ss Tt Uu Vv Ww Xx
Yy Zz 0 1 2 3 4 5
6 7 8 9

All questions asked by five watched experts amaze the judge.
Bodoni MT

All questions asked by five watched experts amaze the judge.
Bodoni MT Bold

All questions asked by five watched experts amaze the judge.
Bodoni MT Italic

All questions asked by five watched experts amaze the judge.
Bodoni MT Bold Italic

Franklin Gothic designed by Morris Fuller Benton

Franklin Gothic has several weights, making it a versatile sans serif choice. The sturdy Franklin Gothic Book is less bold than its Medium counterpart and is nice for small labels.

Country 16pt
STATE 12pt
Major City 12pt
Minor City 10pt
Town 8pt

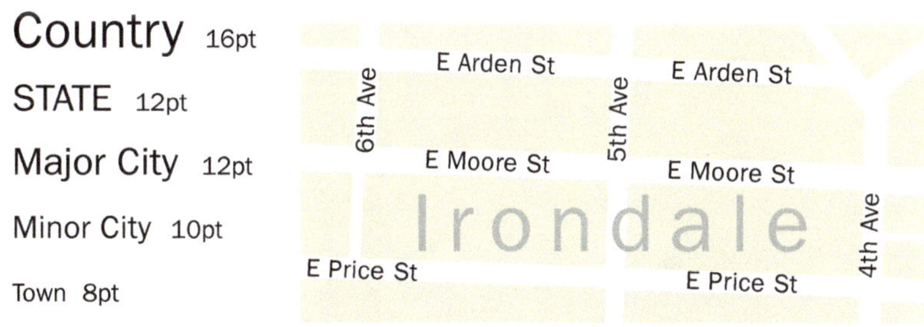

Aa Bb Cc Dd Ee Ff Gg Hh
Ii Jj Kk Ll Mm Nn Oo Pp
Qq Rr Ss Tt Uu Vv Ww Xx
Yy Zz 0 1 2 3 4 5
6 7 8 9

All questions asked by five watched experts amaze the judge.
Franklin Gothic Book

All questions asked by five watched experts amaze the judge.
Franklin Gothic Demi

All questions asked by five watched experts amaze the judge.
Franklin Gothic Book Italic

All questions asked by five watched experts amaze the judge.
Franklin Gothic Demi Cond

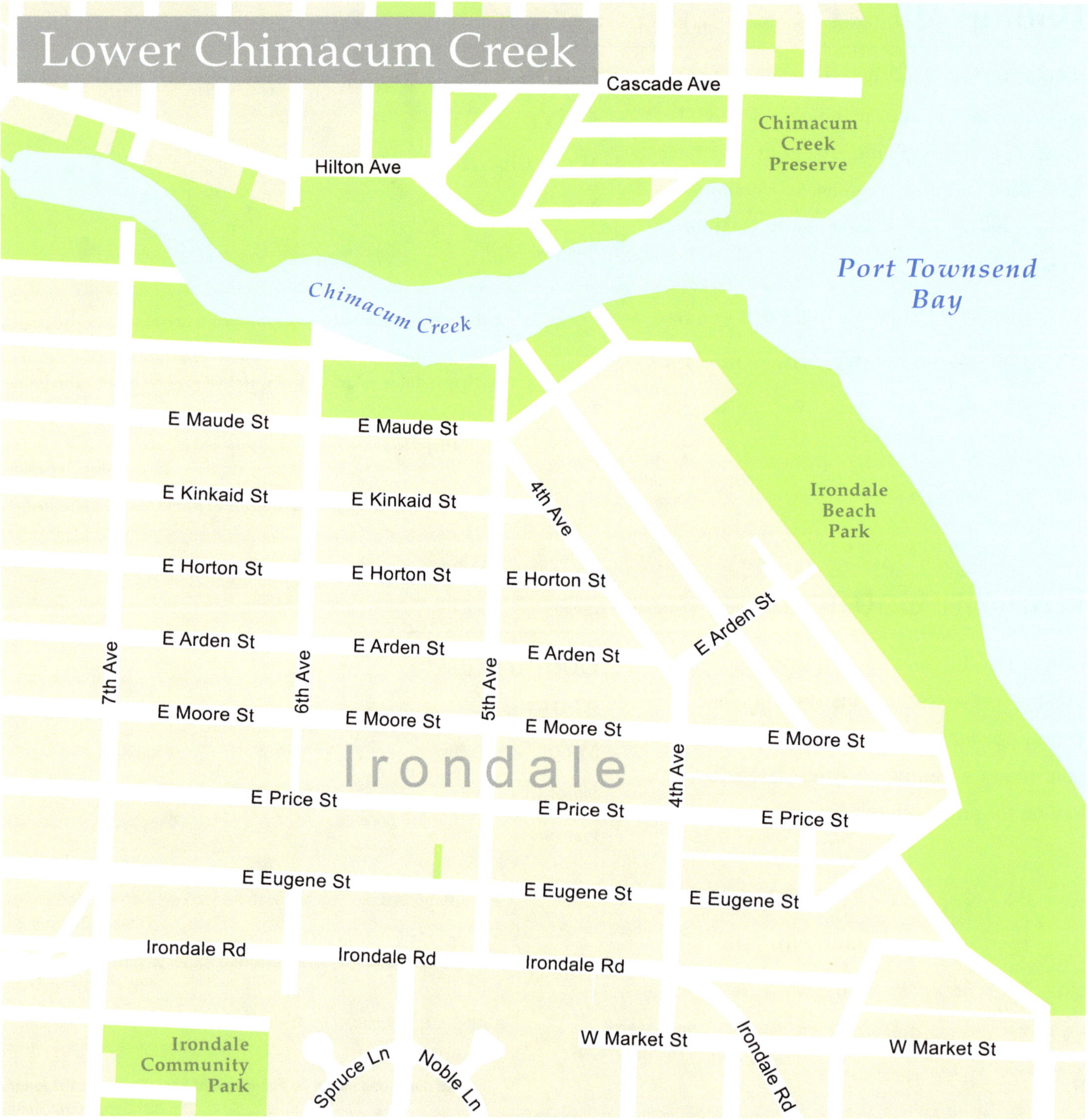

Lower Chimacum Creek

Cascade Ave
Chimacum Creek Preserve
Hilton Ave
Port Townsend Bay
Chimacum Creek
E Maude St
E Maude St
E Kinkaid St
E Kinkaid St
4th Ave
Irondale Beach Park
E Horton St
E Horton St
E Horton St
E Arden St
E Arden St
E Arden St
E Arden St
7th Ave
6th Ave
5th Ave
E Moore St
E Moore St
E Moore St
E Moore St
Irondale
4th Ave
E Price St
E Price St
E Price St
E Eugene St
E Eugene St
E Eugene St
Irondale Rd
Irondale Rd
Irondale Rd
W Market St
W Market St
Irondale Rd
Irondale Community Park
Spruce Ln
Noble Ln

Palatino designed by Hermann Zapf

Palatino is a calligraphy inspired font. It was created to provide maximum legibility on sub-optimal paper types. This makes it ideal for creating readable labels that are layered over complex map images.

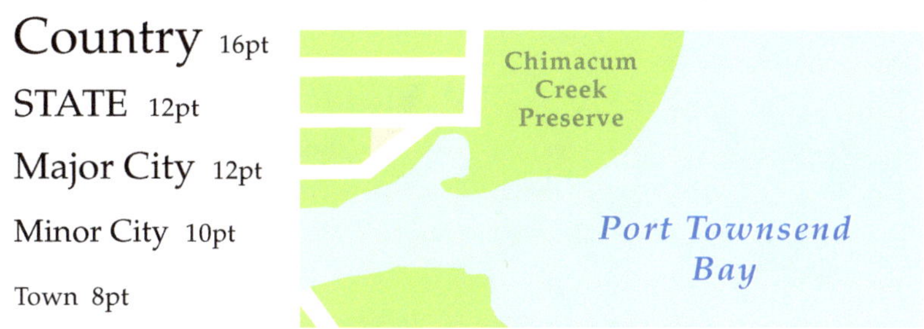

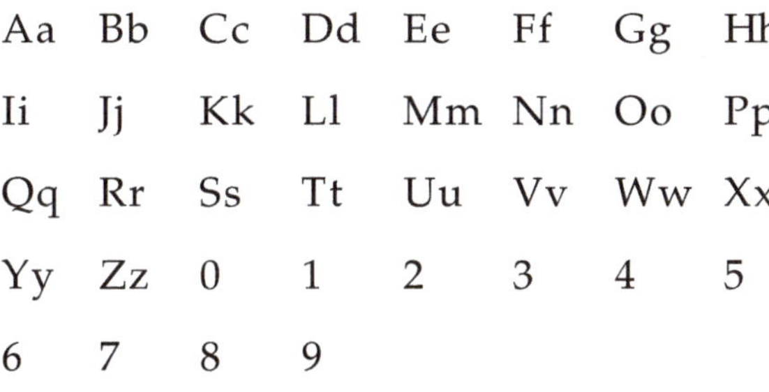

All questions asked by five watched experts amaze the judge.
Palatino

All questions asked by five watched experts amaze the judge.
Palatino Bold

All questions asked by five watched experts amaze the judge.
Palatino Italic

All questions asked by five watched experts amaze the judge.
Palatino Bold Italic

Arial designed by Robin Nicholas and Patricia Saunders

Similar to Helvetica in most glyphs except a, G, Q, R, and 1, Arial is highly readable and was once ubiquitous in print and on the web. It is a fine choice if over-use is not a concern.

Country 16pt
STATE 12pt
Major City 12pt
Minor City 10pt
Town 8pt

E Arden St
E Arden St
6th Ave
5th Ave
E Moore St
E Moore St
Irondale
4th Ave
E Price St
E Price St

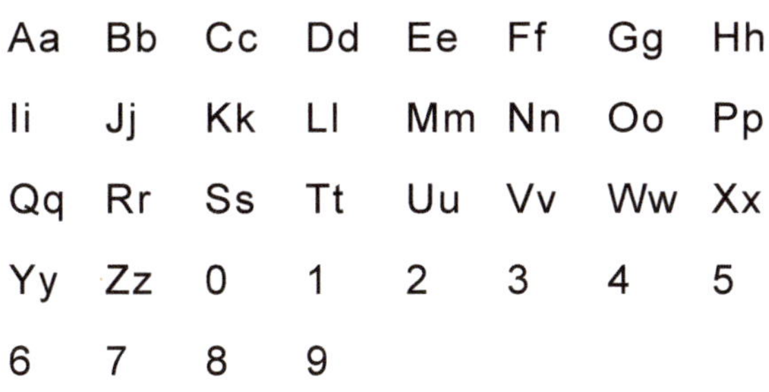

All questions asked by five watched experts amaze the judge.
Arial

All questions asked by five watched experts amaze the judge.
Arial Bold

All questions asked by five watched experts amaze the judge.
Arial Italic

All questions asked by five watched experts amaze the judge.
Arial Bold Italic

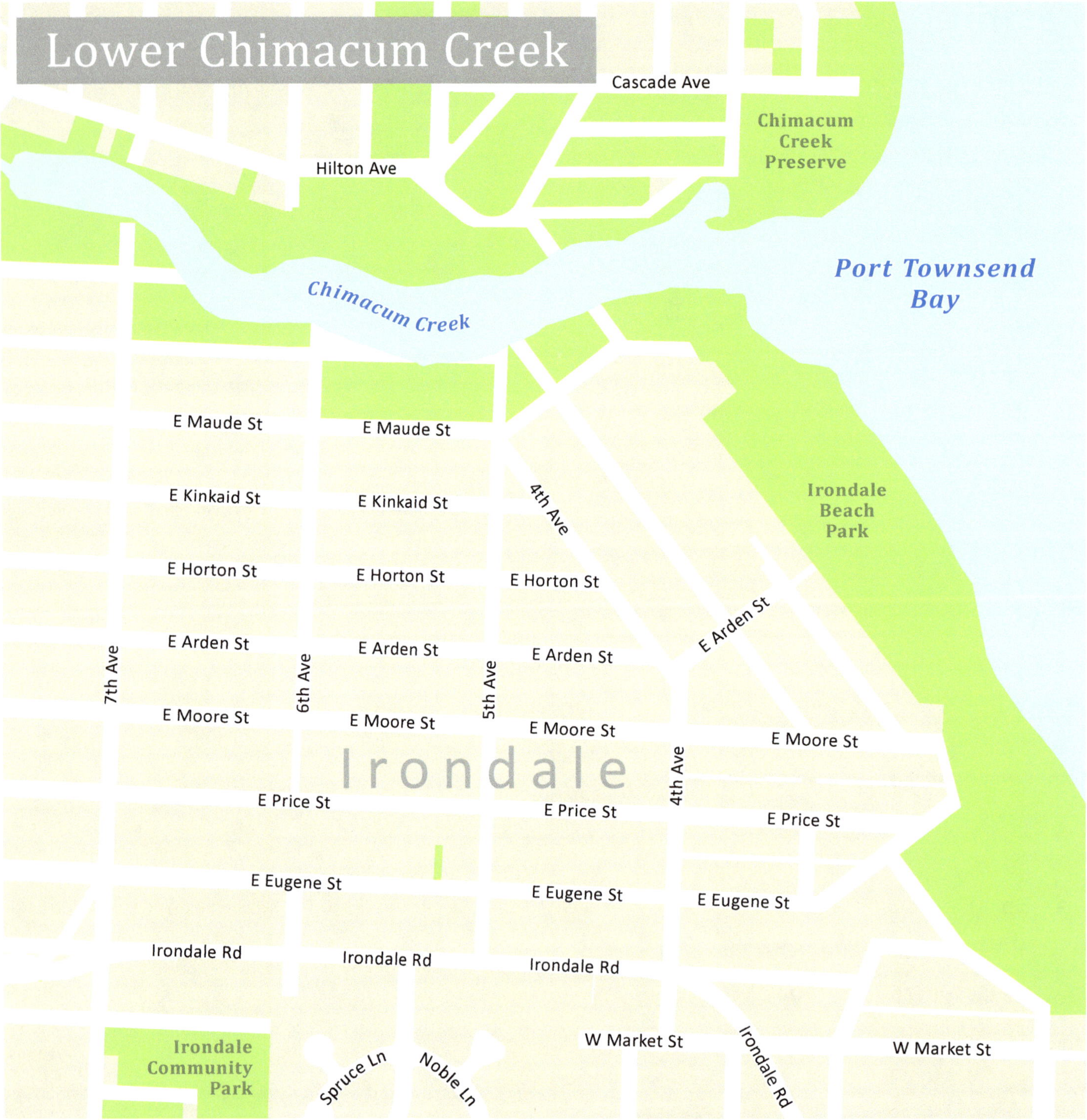

Lower Chimacum Creek

Cascade Ave
Chimacum Creek Preserve
Hilton Ave
Port Townsend Bay
Chimacum Creek
E Maude St
E Maude St
E Kinkaid St
E Kinkaid St
4th Ave
Irondale Beach Park
E Horton St
E Horton St
E Horton St
E Arden St
E Arden St
E Arden St
E Arden St
7th Ave
6th Ave
5th Ave
E Moore St
E Moore St
E Moore St
E Moore St
Irondale
4th Ave
E Price St
E Price St
E Price St
E Eugene St
E Eugene St
E Eugene St
Irondale Rd
Irondale Rd
Irondale Rd
W Market St
W Market St
Irondale Community Park
Spruce Ln
Noble Ln
Irondale Rd

Cambria designed by Jelle Bosma with Steve Matteson and Robin Nicholas

Cambria was designed for on-screen reading and therefore looks good at small sizes. It possesses characteristics of both serif and sans serif. At sizes larger than 20 pts, reduce the tracking for a better result.

Country 16pt

STATE 12pt

Major City 12pt

Minor City 10pt

Town 8pt

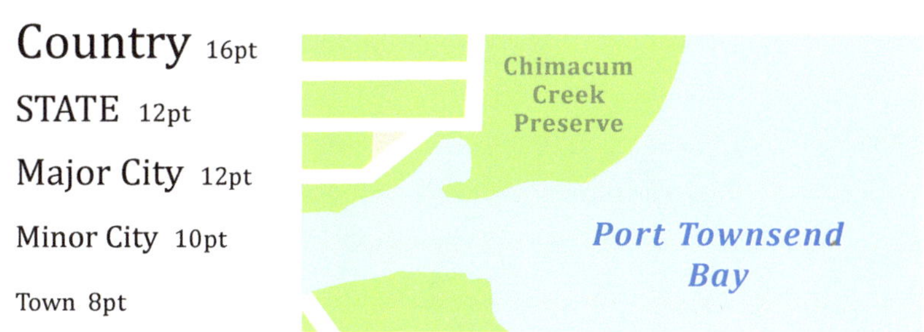

Aa Bb Cc Dd Ee Ff Gg Hh

Ii Jj Kk Ll Mm Nn Oo Pp

Qq Rr Ss Tt Uu Vv Ww Xx

Yy Zz 0 1 2 3 4 5

6 7 8 9

All questions asked by five watched experts amaze the judge.
Cambria

All questions asked by five watched experts amaze the judge.
Cambria Bold

All questions asked by five watched experts amaze the judge.
Cambria Italic

All questions asked by five watched experts amaze the judge.
Cambria Bold Italic

Calibri designed by Lucas de Groot

The slightly rounded corners on the glyphs in this business minded typeface allow it to take on a more easy-going attitude than it would otherwise. It is a great combination of serious and friendly.

Country 16pt

STATE 12pt

Major City 12pt

Minor City 10pt

Town 8pt

E Arden St
E Arden St
6th Ave
5th Ave
E Moore St
E Moore St
Irondale
4th Ave
E Price St
E Price St

Aa Bb Cc Dd Ee Ff Gg Hh

Ii Jj Kk Ll Mm Nn Oo Pp

Qq Rr Ss Tt Uu Vv Ww Xx

Yy Zz 0 1 2 3 4 5

6 7 8 9

All questions asked by five watched experts amaze the judge.
Calibri

All questions asked by five watched experts amaze the judge.
Calibri Bold

All questions asked by five watched experts amaze the judge.
Calibri Italic

All questions asked by five watched experts amaze the judge.
Calibri Bold Italic

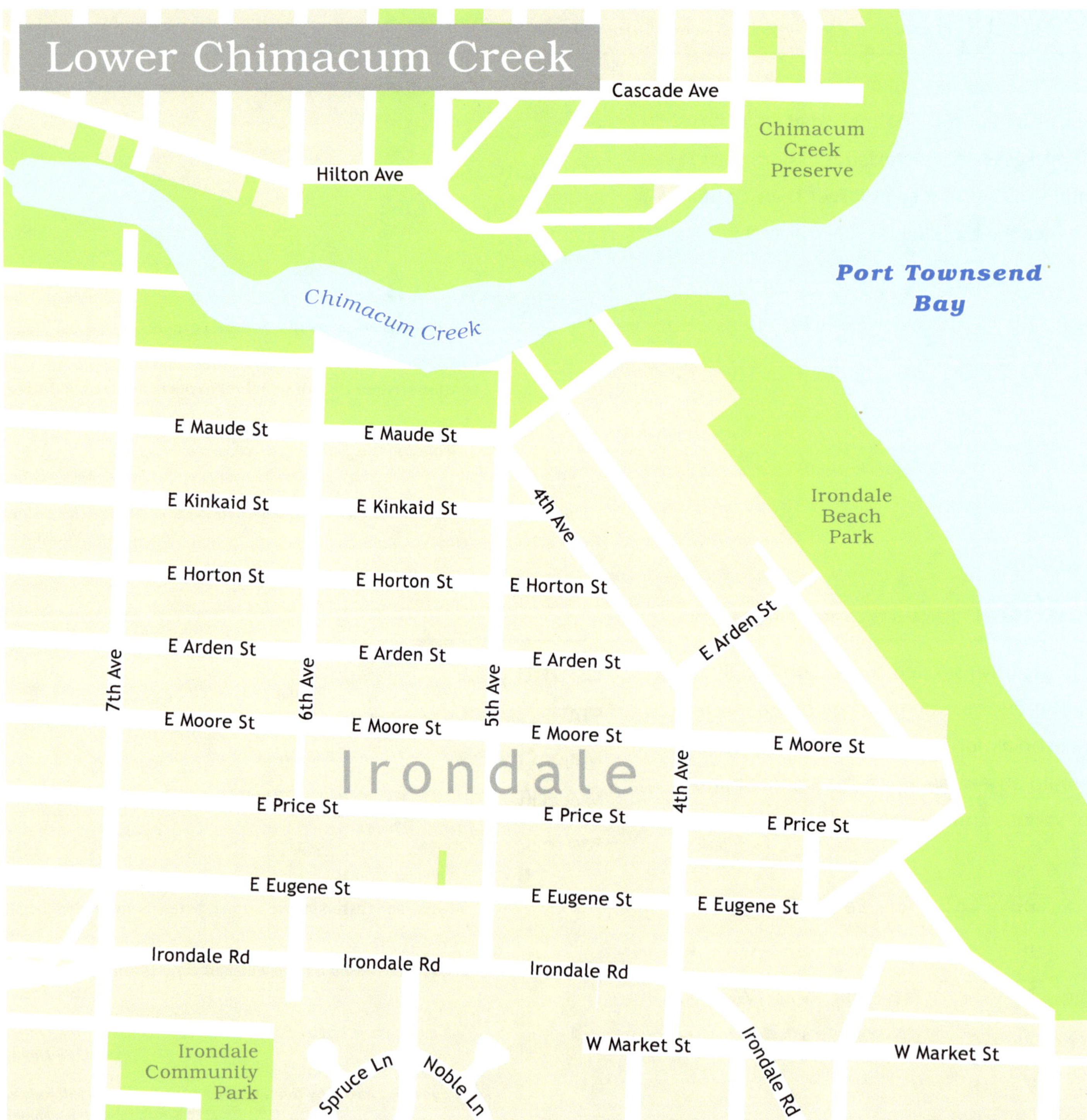
Lower Chimacum Creek
Cascade Ave
Hilton Ave
Chimacum Creek Preserve
Port Townsend Bay
Chimacum Creek
E Maude St
E Maude St
E Kinkaid St
E Kinkaid St
4th Ave
Irondale Beach Park
E Horton St
E Horton St
E Horton St
E Arden St
E Arden St
E Arden St
E Arden St
E Arden St
7th Ave
6th Ave
5th Ave
E Moore St
E Moore St
E Moore St
E Moore St
Irondale
4th Ave
E Price St
E Price St
E Price St
E Eugene St
E Eugene St
E Eugene St
Irondale Rd
Irondale Rd
Irondale Rd
W Market St
W Market St
Irondale Rd
Irondale Community Park
Spruce Ln
Noble Ln

Bookman designed by Ong Chong Wah

Distributed as Bookman Old Style in Microsoft® Office products, this font is good for titles while also readable at small sizes due to a tall x-height. It supplies text with an authoritative personality.

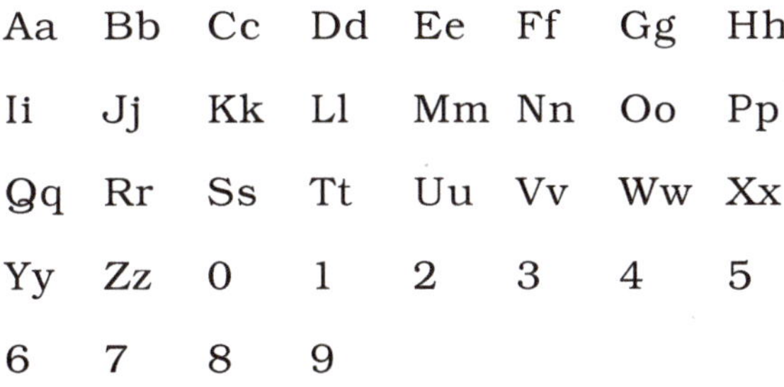
Aa Bb Cc Dd Ee Ff Gg Hh
Ii Jj Kk Ll Mm Nn Oo Pp
Qq Rr Ss Tt Uu Vv Ww Xx
Yy Zz 0 1 2 3 4 5
6 7 8 9

Country 16pt

STATE 12pt

Major City 12pt

Minor City 10pt

Town 8pt

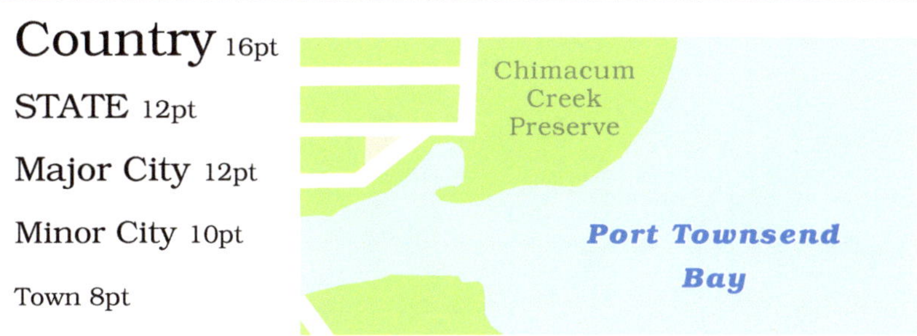

All questions asked by five watched experts amaze the judge.
Bookman Light

All questions asked by five watched experts amaze the judge.
Bookman Bold

All questions asked by five watched experts amaze the judge.
Bookman Light Italic

All questions asked by five watched experts amaze the judge.
Bookman Bold Italic

Trebuchet MS designed by Vincent Connare

A common web font, Trebuchet MS has a tall x-height and blocky strokes. It is especially well suited for large headings and map titles while remaining easy to read at small sizes.

Aa Bb Cc Dd Ee Ff Gg Hh
Ii Jj Kk Ll Mm Nn Oo Pp
Qq Rr Ss Tt Uu Vv Ww Xx
Yy Zz 0 1 2 3 4 5
6 7 8 9

Country 16pt

STATE 12pt

Major City 12pt

Minor City 10pt

Town 8pt

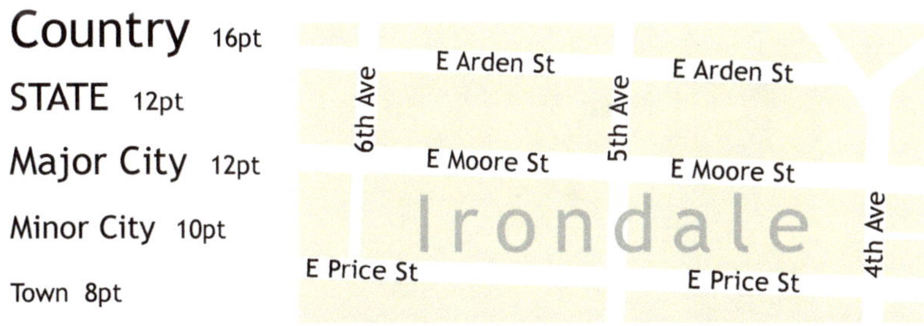

All questions asked by five watched experts amaze the judge.
Trebuchet MS

All questions asked by five watched experts amaze the judge.
Trebuchet MS Bold

All questions asked by five watched experts amaze the judge.
Trebuchet MS Italic

All questions asked by five watched experts amaze the judge.
Trebuchet MS Bold Italic

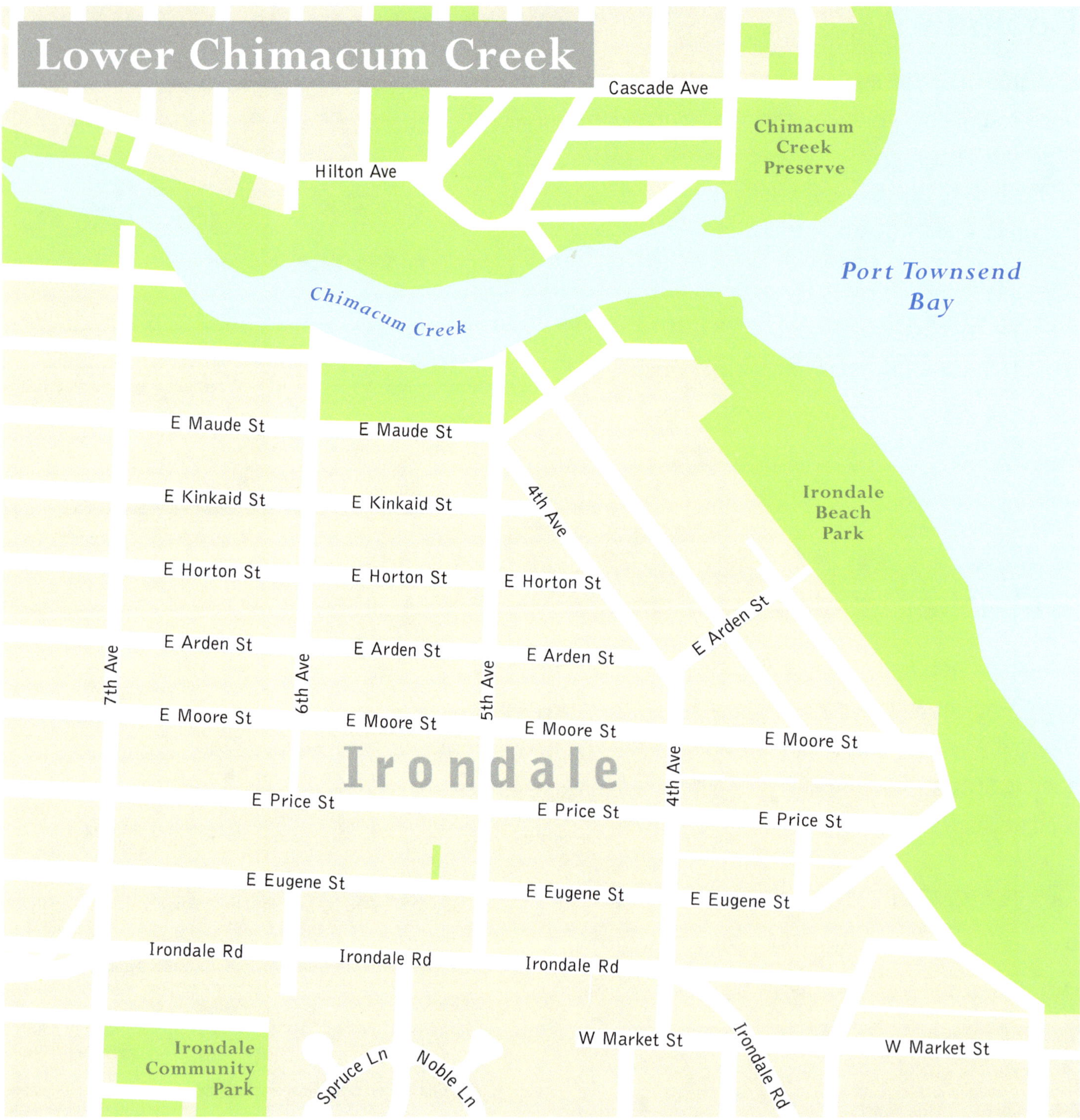
Lower Chimacum Creek
Cascade Ave
Hilton Ave
Chimacum Creek Preserve
Chimacum Creek
Port Townsend Bay
E Maude St
E Maude St
E Kinkaid St
E Kinkaid St
4th Ave
Irondale Beach Park
E Horton St
E Horton St
E Horton St
E Arden St
E Arden St
E Arden St
E Arden St
E Arden St
7th Ave
6th Ave
5th Ave
E Moore St
E Moore St
E Moore St
E Moore St
Irondale
4th Ave
E Price St
E Price St
E Price St
E Eugene St
E Eugene St
E Eugene St
Irondale Rd
Irondale Rd
Irondale Rd
W Market St
W Market St
Irondale Rd
Irondale Community Park
Spruce Ln
Noble Ln

Perpetua designed by Eric Gill

Perpetua is mostly a straight-laced typeface but does have some touches of informality. Its chiseled look and small, diagonal serifs give it authority. A good choice for historic mapping or other formal map products.

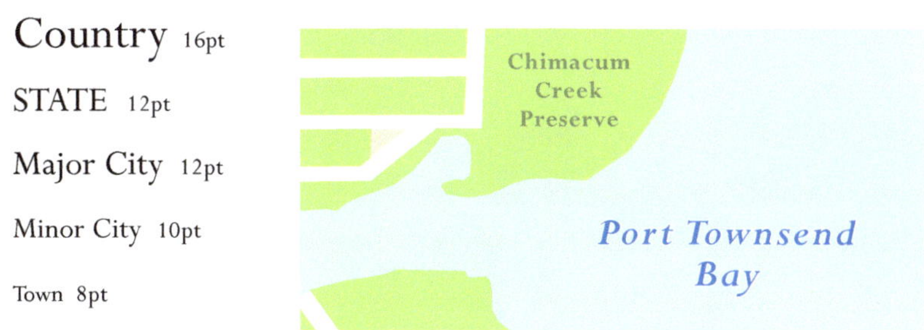

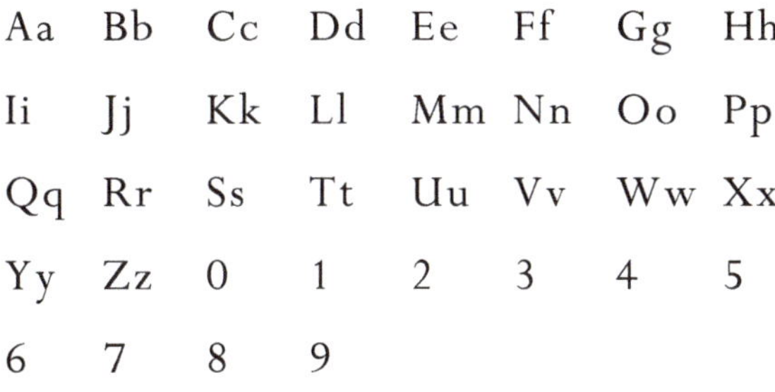

All questions asked by five watched experts amaze the judge.
Perpetua

All questions asked by five watched experts amaze the judge.
Perpetua Bold

All questions asked by five watched experts amaze the judge.
Perpetua Italic

All questions asked by five watched experts amaze the judge.
Perpetua Bold Italic

Bell Gothic designed by Chauncey H. Griffith

Bell Gothic was originally created for Bell telephone directories and is thus very space efficient. It has a dynamic and showy personality. Bell Gothic Light (shown here) is great for data tables and labels.

Country 16pt

STATE 12pt

Major City 12pt

Minor City 10pt

Town 8pt

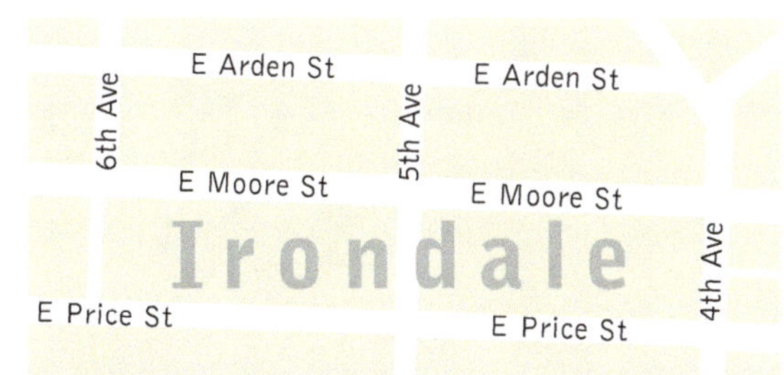

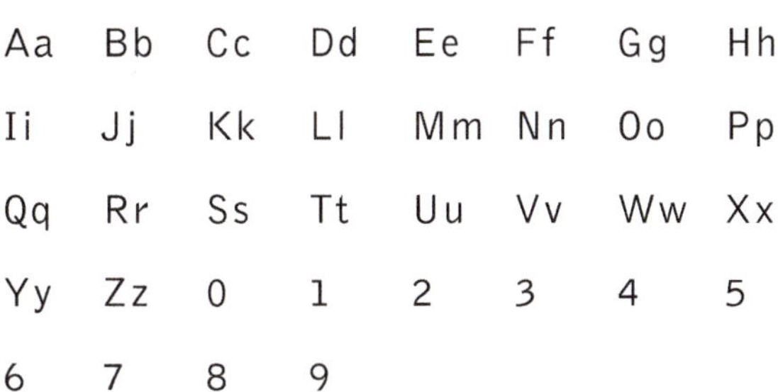

All questions asked by five watched experts amaze the judge.
Bell Gothic Std Light

All questions asked by five watched experts amaze the judge.
Black Gothic Std Black

All questions asked by five watched experts amaze the judge.
Bell Gothic Italic

All questions asked by five watched experts amaze the judge.
Bell Gothic Bold Italic

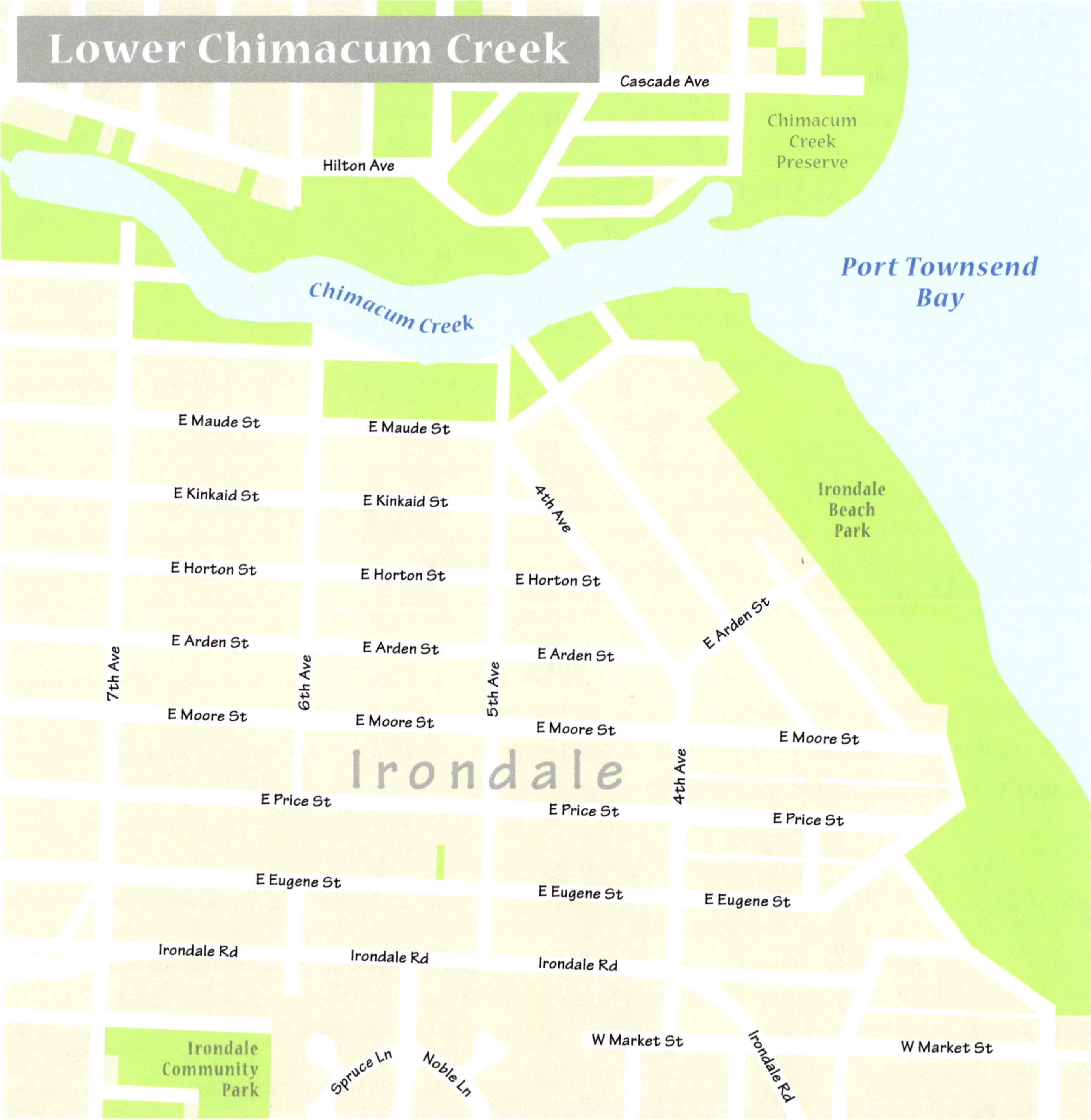

Lower Chimacum Creek

Cascade Ave
Chimacum Creek Preserve
Hilton Ave
Port Townsend Bay
Chimacum Creek
E Maude St
E Maude St
E Kinkaid St
E Kinkaid St
4th Ave
Irondale Beach Park
E Horton St
E Horton St
E Horton St
E Arden St
E Arden St
E Arden St
E Arden St
7th Ave
6th Ave
5th Ave
E Moore St
E Moore St
E Moore St
E Moore St
Irondale
4th Ave
E Price St
E Price St
E Price St
E Eugene St
E Eugene St
E Eugene St
Irondale Rd
Irondale Rd
Irondale Rd
W Market St
W Market St
Irondale Rd
Irondale Community Park
Spruce Ln
Noble Ln

Nueva designed by Carol Twombly

Nueva has a high stroke contrast, making it a very unique typeface with lively glyphs. It has a true italic version that is effective for water labeling. It comes in many weights including light, condensed (shown here), regular and bold. Overall, this typeface has an informal and friendly feel.

Country 16pt

STATE 12pt

Major City 12pt

Minor City 10pt

Town 8pt

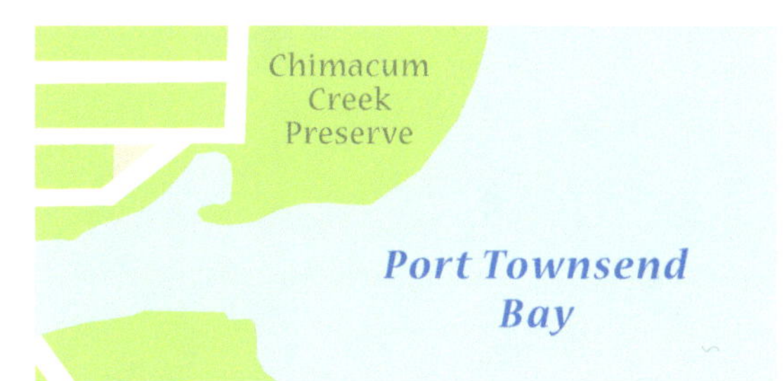

Aa Bb Cc Dd Ee Ff Gg Hh

Ii Jj Kk Ll Mm Nn Oo Pp

Qq Rr Ss Tt Uu Vv Ww Xx

Yy Zz 0 1 2 3 4 5

6 7 8 9

All questions asked by five watched experts amaze the judge.
Nueva Std Condensed

All questions asked by five watched experts amaze the judge.
Nueva Std Condensed Bold

All questions asked by five watched experts amaze the judge.
Nueva Std Condensed Italic

All questions asked by five watched experts amaze the judge.
Nueva Std Condensed Bold Italic

Tekton Pro designed Jim Wasco, David Siegel, Christopher Slye

Tekton Pro is based on architect hand lettering, complete with upward slants on many of the letters. Informal maps and those based on architectural plans might use this typeface.

Country 16pt

STATE 12pt

Major City 12pt

Minor City 10pt

Town 8pt

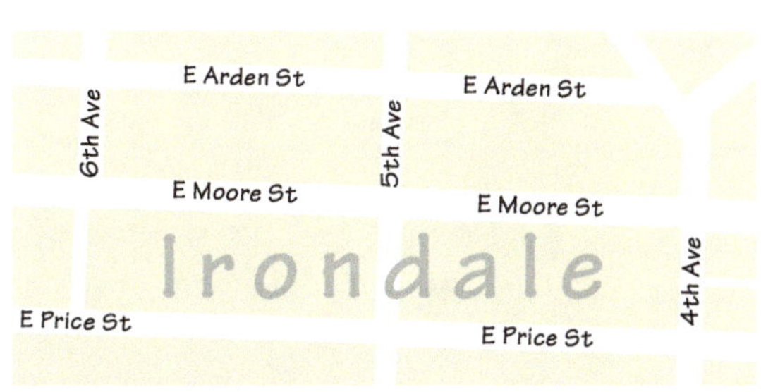

Aa Bb Cc Dd Ee Ff Gg Hh

Ii Jj Kk Ll Mm Nn Oo Pp

Qq Rr Ss Tt Uu Vv Ww Xx

Yy Zz 0 1 2 3 4 5

6 7 8 9

All questions asked by five watched experts amaze the judge.
Tekton Pro Bold

All questions asked by five watched experts amaze the judge.
Tekton Pro Bold Oblique

All questions asked by five watched experts amaze the judge.
Tekton Pro Ext

All questions asked by five watched experts amaze the judge.
Tekton Pro Cond

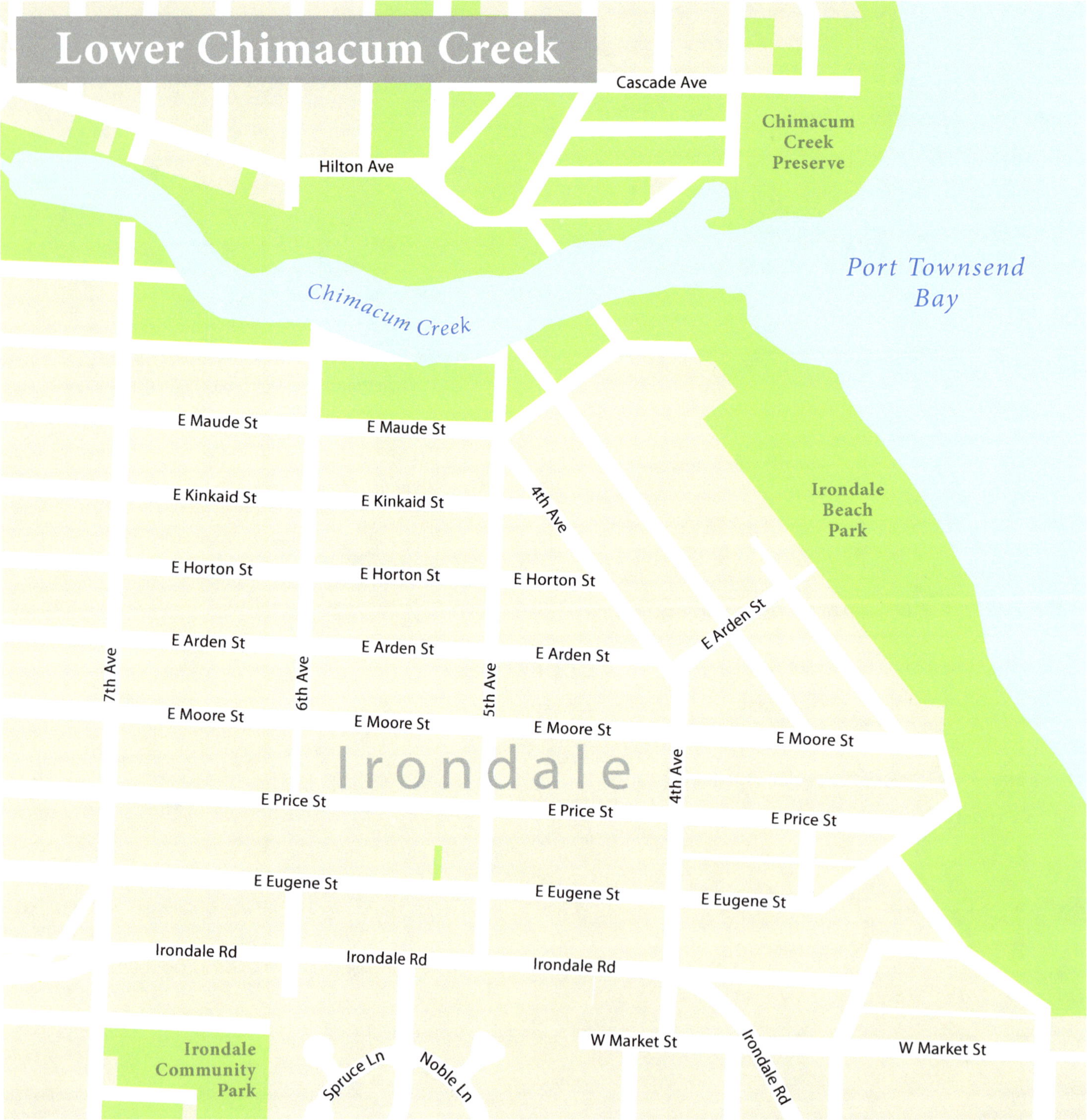

Lower Chimacum Creek

Cascade Ave
Chimacum Creek Preserve
Hilton Ave
Chimacum Creek
Port Townsend Bay
E Maude St
E Maude St
E Kinkaid St
E Kinkaid St
4th Ave
Irondale Beach Park
E Horton St
E Horton St
E Horton St
E Arden St
E Arden St
E Arden St
E Arden St
7th Ave
6th Ave
5th Ave
E Moore St
E Moore St
E Moore St
E Moore St
Irondale
4th Ave
E Price St
E Price St
E Price St
E Eugene St
E Eugene St
E Eugene St
Irondale Rd
Irondale Rd
Irondale Rd
W Market St
W Market St
Irondale Rd
Irondale Community Park
Spruce Ln
Noble Ln

Minion designed by Robert Slimbach

Reminiscent of typefaces from the late Renaissance, Minion is a great font for use in large bodies of text and for authoritative looking titles. It has a great variety of weights and forms, giving it good potential for complex labeling hierarchies.

Country 16pt

STATE 12pt

Major City 12pt

Minor City 10pt

Town 8pt

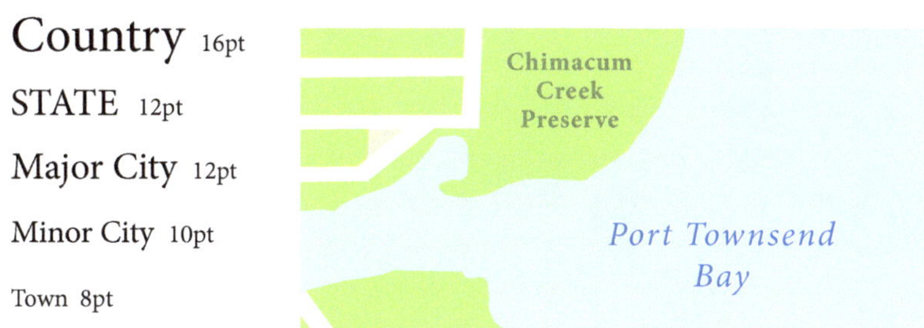

Aa Bb Cc Dd Ee Ff Gg Hh

Ii Jj Kk Ll Mm Nn Oo Pp

Qq Rr Ss Tt Uu Vv Ww Xx

Yy Zz 0 1 2 3 4 5

6 7 8 9

All questions asked by five watched experts amaze the judge.
Minion Pro

All questions asked by five watched experts amaze the judge.
Minion Pro Bold

All questions asked by five watched experts amaze the judge.
Minion Pro Italic

All questions asked by five watched experts amaze the judge.
Minion Pro Bold Italic

Myriad Pro designed by Robert Slimbach and Carol Twombly

Myriad Pro has a clean, easy feel. Its warmth, along with its readability, have made it a popular contemporary font. It is a good choice for labels that aspire to be read rather than looked at.

Country 16pt

STATE 12pt

Major City 12pt

Minor City 10pt

Town 8pt

E Arden St
E Arden St
6th Ave
5th Ave
E Moore St
E Moore St
Irondale
4th Ave
E Price St
E Price St

Aa Bb Cc Dd Ee Ff Gg Hh

Ii Jj Kk Ll Mm Nn Oo Pp

Qq Rr Ss Tt Uu Vv Ww Xx

Yy Zz 0 1 2 3 4 5

6 7 8 9

All questions asked by five watched experts amaze the judge.
Myriad Pro

All questions asked by five watched experts amaze the judge.
Myriad Pro Bold

All questions asked by five watched experts amaze the judge.
Myriad Pro Italic

All questions asked by five watched experts amaze the judge.
Myriad Pro Bold Italic

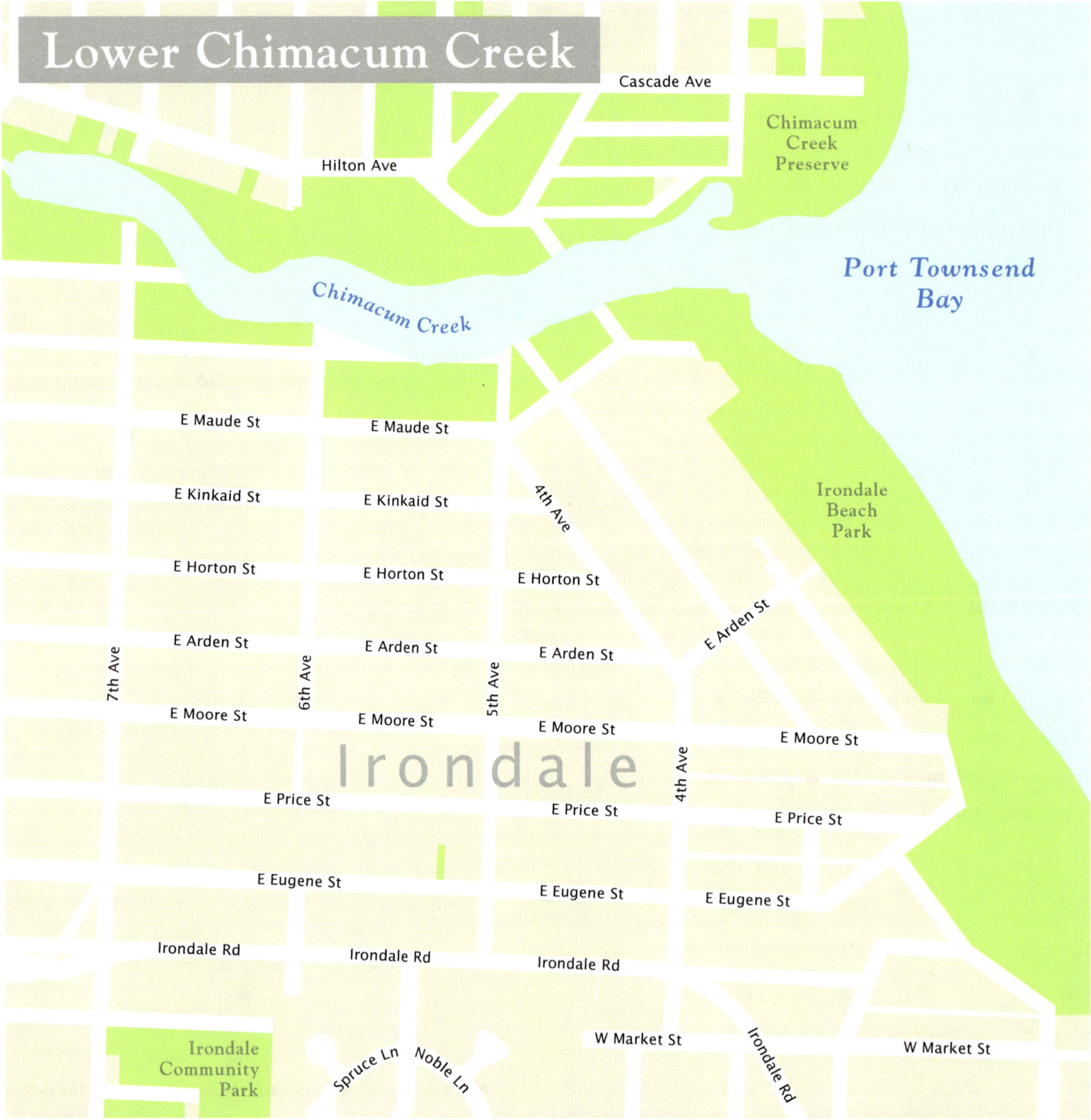

Lower Chimacum Creek

Cascade Ave
Chimacum Creek Preserve
Hilton Ave
Chimacum Creek
Port Townsend Bay
E Maude St
E Maude St
E Kinkaid St
E Kinkaid St
4th Ave
Irondale Beach Park
E Horton St
E Horton St
E Horton St
E Arden St
E Arden St
E Arden St
E Arden St
7th Ave
6th Ave
5th Ave
E Moore St
E Moore St
E Moore St
E Moore St
Irondale
4th Ave
E Price St
E Price St
E Price St
E Eugene St
E Eugene St
E Eugene St
Irondale Rd
Irondale Rd
Irondale Rd
W Market St
W Market St
Irondale Rd
Irondale Community Park
Spruce Ln
Noble Ln

Goudy Old Style designed by Frederic Goudy

Goudy Old Style is very legible and readable in print and may be the most popular typeface ever made. While it is nice in larger blocks of text, it is a standard for larger type as well, making it a friendly and non-obtrusive titling typeface.

Country 16pt

STATE 12pt

Major City 12pt

Minor City 10pt

Town 8pt

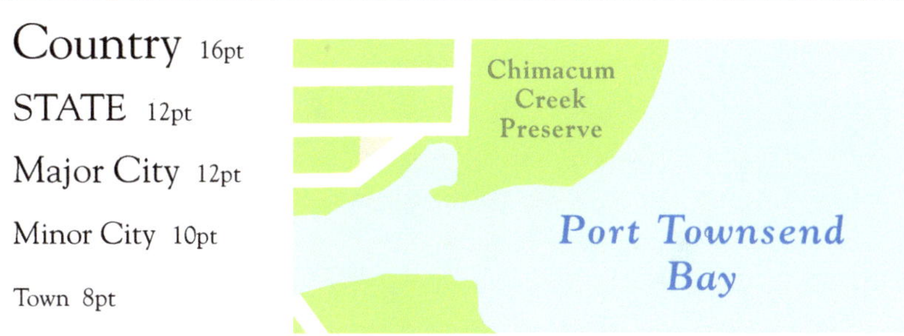

Aa Bb Cc Dd Ee Ff Gg Hh

Ii Jj Kk Ll Mm Nn Oo Pp

Qq Rr Ss Tt Uu Vv Ww Xx

Yy Zz 0 1 2 3 4 5

6 7 8 9

All questions asked by five watched experts amaze the judge.
Goudy Old Style

All questions asked by five watched experts amaze the judge.
Goudy Old Style Bold

All questions asked by five watched experts amaze the judge.
Goudy Old Style Italic

All questions asked by five watched experts amaze the judge.
Goudy Old Style Bold Italic

Lucida Sans designed by Charles Bigelow and Kris Holmes

A highly versatile and sometimes overlooked typeface, Lucida Sans has a persuasive weightiness. Its calligraphy style italics are especially well-suited for water labeling.

Country 16pt

STATE 12pt

Major City 12pt

Minor City 10pt

Town 8pt

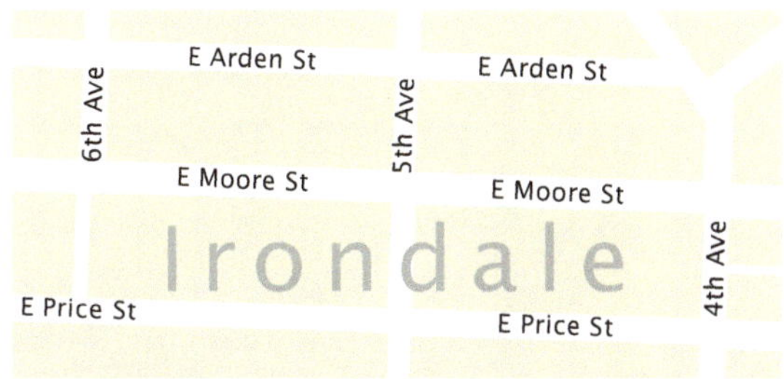

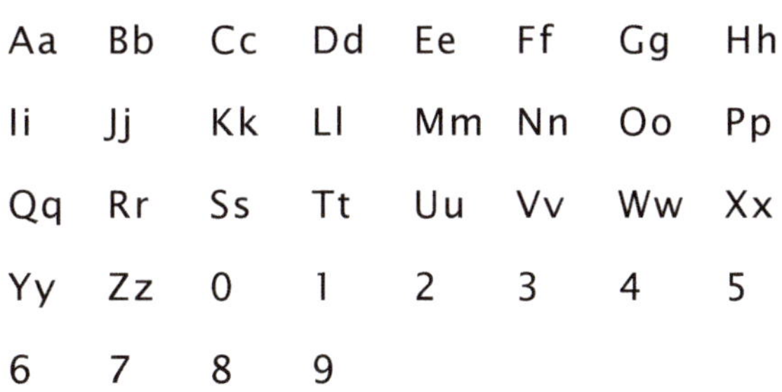

All questions asked by five watched experts amaze the judge.
Lucida Sans

All questions asked by five watched experts amaze the judge.
Lucida Sans Bold

All questions asked by five watched experts amaze the judge.
Lucida Sans Italic

All questions asked by five watched experts amaze the judge.
Lucida Sans Bold Italic

FREE TYPEFACES

These typefaces are free and currently allow commercial use. As always, however, read their respective end user license agreements when you download them, as they sometimes change. Free typefaces can be risky as many are of poor quality. However, these have been vetted for quality and applicability to map design.

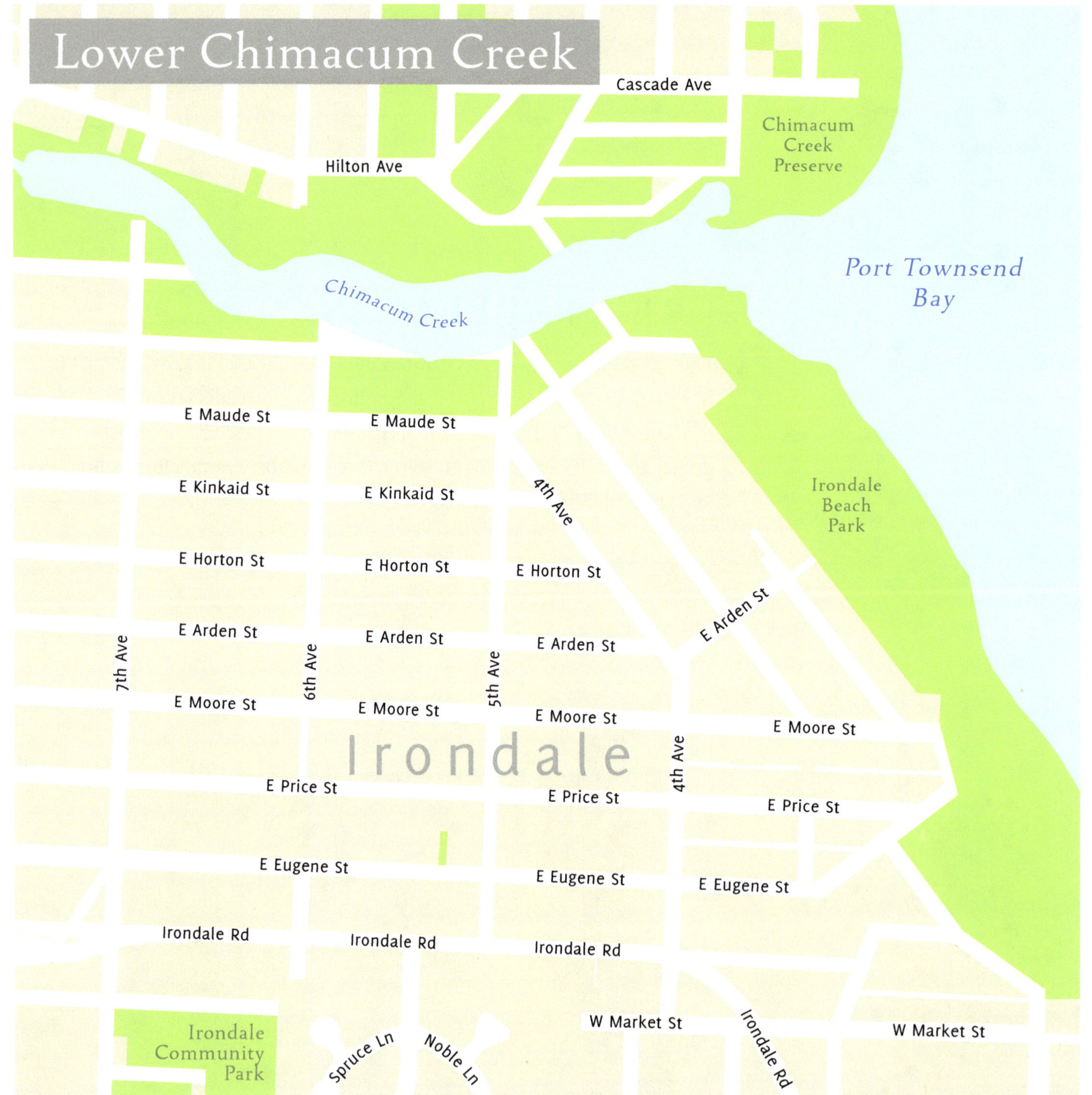
Lower Chimacum Creek
Cascade Ave
Hilton Ave
Chimacum Creek Preserve
Port Townsend Bay
Chimacum Creek
E Maude St
E Maude St
E Kinkaid St
E Kinkaid St
4th Ave
Irondale Beach Park
E Horton St
E Horton St
E Horton St
E Arden St
E Arden St
E Arden St
E Arden St
7th Ave
6th Ave
5th Ave
E Moore St
E Moore St
E Moore St
E Moore St
Irondale
4th Ave
E Price St
E Price St
E Price St
E Eugene St
E Eugene St
E Eugene St
Irondale Rd
Irondale Rd
Irondale Rd
W Market St
W Market St
Irondale Rd
Irondale Community Park
Spruce Ln
Noble Ln

Tallys designed by Jos Buivenga

Tallys has a distinctive, fluid, style that lends credence to map titles, legends, explanatory text, and natural feature labels. While it only has one weight and small x-heights, it is still a unique and clean display typeface.

Country 16pt

STATE 12pt

Major City 12pt

Minor City 10pt

Town 8pt

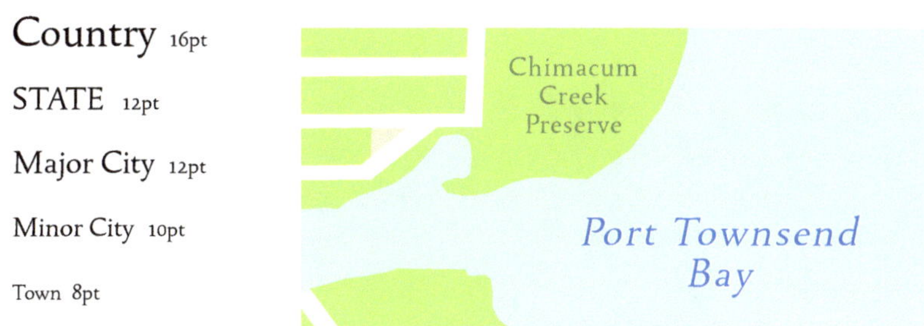

Aa	Bb	Cc	Dd	Ee	Ff	Gg	Hh
Ii	Jj	Kk	Ll	Mm	Nn	Oo	Pp
Qq	Rr	Ss	Tt	Uu	Vv	Ww	Xx
Yy	Zz	0	1	2	3	4	5
6	7	8	9				

All questions asked by five watched experts amaze the judge.
Tallys

Delicious designed by Jos Buivenga

Delicious is an amusing font with quirky flourishes (e.g., the slight bend in the letters l and e). It has a large x-height. With an exquisite true italic and a nice small cap variety, it provides many options.

Country 16pt

STATE 12pt

Major City 12pt

Minor City 10pt

Town 8pt

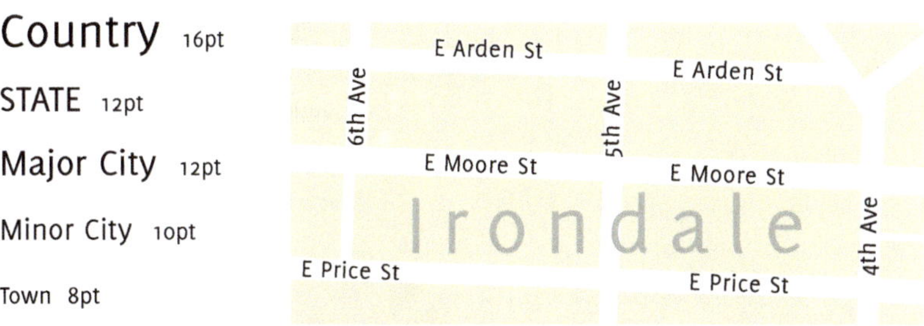

Aa	Bb	Cc	Dd	Ee	Ff	Gg	Hh
Ii	Jj	Kk	Ll	Mm	Nn	Oo	Pp
Qq	Rr	Ss	Tt	Uu	Vv	Ww	Xx
Yy	Zz	0	1	2	3	4	5
6	7	8	9				

All questions asked by five watched experts amaze the judge.
Delicious

All questions asked by five watched experts amaze the judge.
Delicious Heavy

All questions asked by five watched experts amaze the judge.
Delicious Italic

ALL QUESTIONS ASKED BY FIVE WATCHED EXPERTS AMAZE THE JUDGE.
Delicious Small Caps

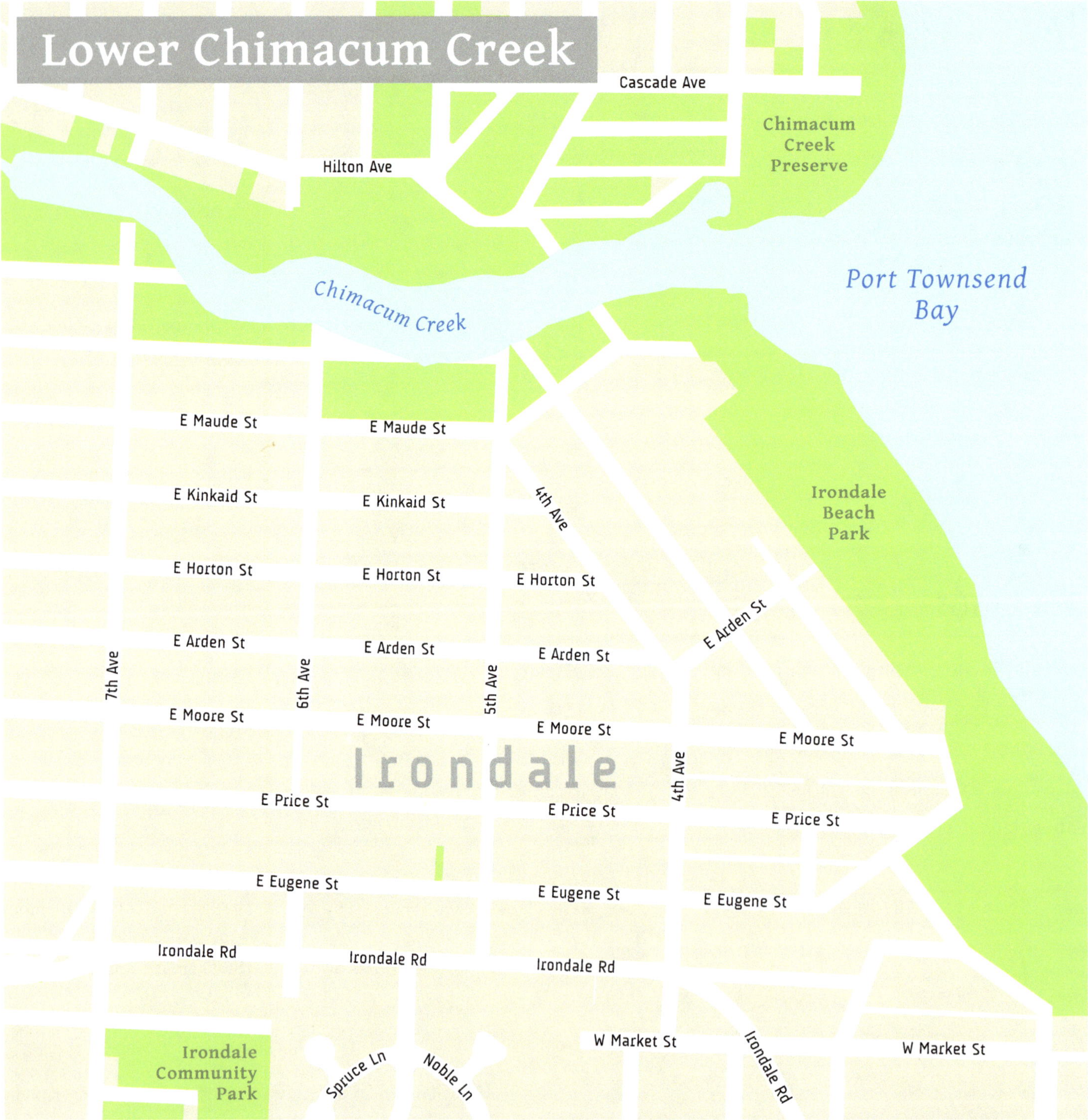

Lower Chimacum Creek

Cascade Ave
Chimacum Creek Preserve
Hilton Ave
Chimacum Creek
Port Townsend Bay
E Maude St
E Maude St
E Kinkaid St
E Kinkaid St
4th Ave
Irondale Beach Park
E Horton St
E Horton St
E Horton St
E Arden St
E Arden St
E Arden St
E Arden St
7th Ave
6th Ave
5th Ave
E Moore St
E Moore St
E Moore St
E Moore St
Irondale
4th Ave
E Price St
E Price St
E Price St
E Eugene St
E Eugene St
E Eugene St
Irondale Rd
Irondale Rd
Irondale Rd
W Market St
W Market St
Irondale Rd
Irondale Community Park
Spruce Ln
Noble Ln

Gentium Plus designed by Victor Gaultney

One of the distinguishing features of Gentium Plus is the vast amount of Latin-based alphabets it supports. It is highly readable. Other varieties are Gentium Book, Gentium Basic, and Gentium Compact.

Country 16pt

STATE 12pt

Major City 12pt

Minor City 10pt

Town 8pt

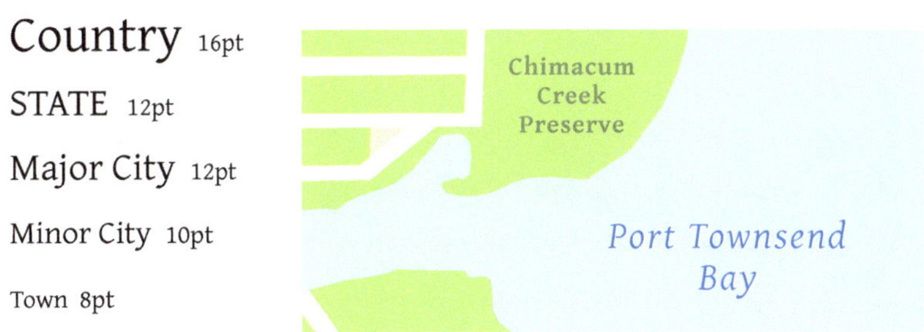

Aa	Bb	Cc	Dd	Ee	Ff	Gg	Hh
Ii	Jj	Kk	Ll	Mm	Nn	Oo	Pp
Qq	Rr	Ss	Tt	Uu	Vv	Ww	Xx
Yy	Zz	0	1	2	3	4	5
6	7	8	9				

All questions asked by five watched experts amaze the judge.
Gentium Plus

All questions asked by five watched experts amaze the judge.
Gentium Plus Italic

All questions asked by five watched experts amaze the judge.
Gentium Book Basic

All questions asked by five watched experts amaze the judge.
Gentium Basic Bold

AUdimat designed by Jack Usine

AUdimat is a modified sans serif/slab serif design with a techie personality. While it may look its best at large sizes, its ability to look at-home as a linear feature labeler is also apparent.

Country 16pt

STATE 12pt

Major City 12pt

Minor City 10pt

Town 8pt

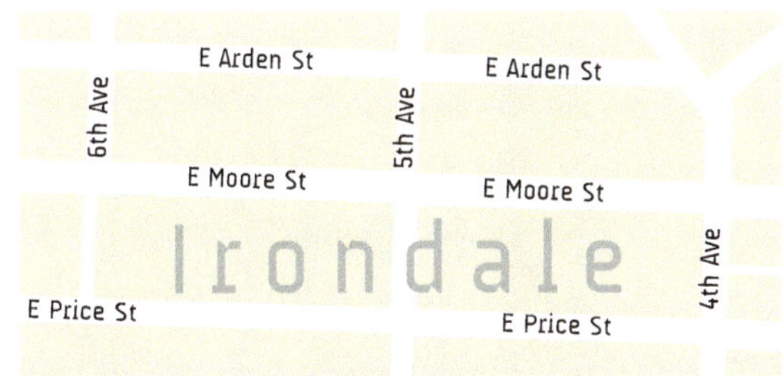

Aa	Bb	Cc	Dd	Ee	Ff	Gg	Hh
Ii	Jj	Kk	Ll	Mm	Nn	Oo	Pp
Qq	Rr	Ss	Tt	Uu	Vv	Ww	Xx
Yy	Zz	0	1	2	3	4	5
6	7	8	9				

All questions asked by five watched experts amaze the judge.
AUdimat

All questions asked by five watched experts amaze the judge.
AUdimat Bold

All questions asked by five watched experts amaze the judge.
AUdimat Italic

All questions asked by five watched experts amaze the judge.
AUdimat Bold Italic

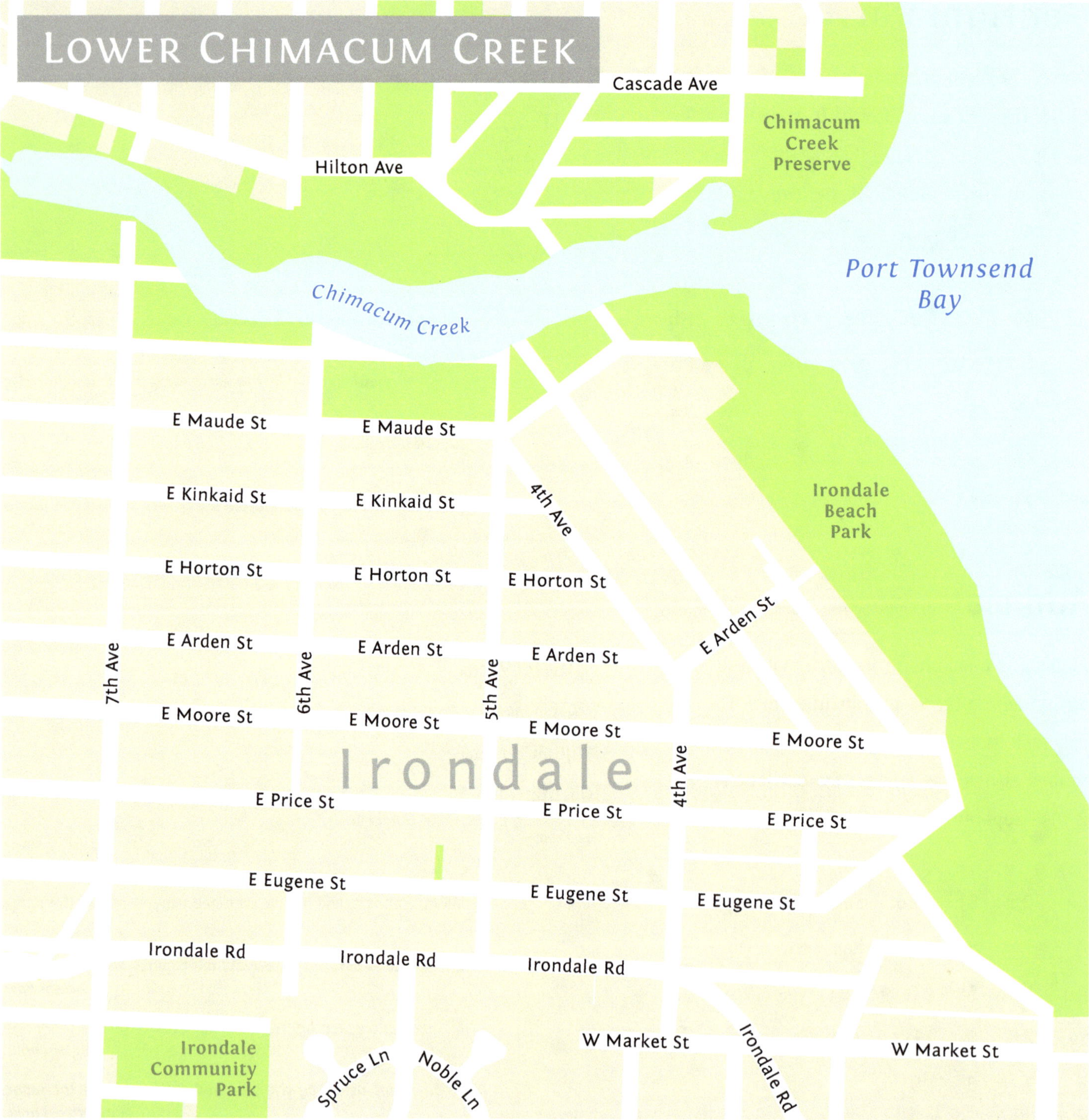

Lower Chimacum Creek

Cascade Ave
Chimacum Creek Preserve
Hilton Ave
Chimacum Creek
Port Townsend Bay
E Maude St
E Maude St
E Kinkaid St
E Kinkaid St
4th Ave
Irondale Beach Park
E Horton St
E Horton St
E Horton St
E Arden St
E Arden St
E Arden St
E Arden St
E Arden St
7th Ave
6th Ave
5th Ave
E Moore St
E Moore St
E Moore St
E Moore St
Irondale
4th Ave
E Price St
E Price St
E Price St
E Eugene St
E Eugene St
E Eugene St
Irondale Rd
Irondale Rd
Irondale Rd
W Market St
W Market St
Irondale Rd
Irondale Community Park
Spruce Ln
Noble Ln

Fontin designed by Jos Buivenga

Fontin is a nice, dark font that works well in small text sizes due to its tall x-height. With friendly curves balanced by a heavier weight, its personality is of both grace and importance.

Country 16pt

STATE 12pt

Major City 12pt

Minor City 10pt

Town 8pt

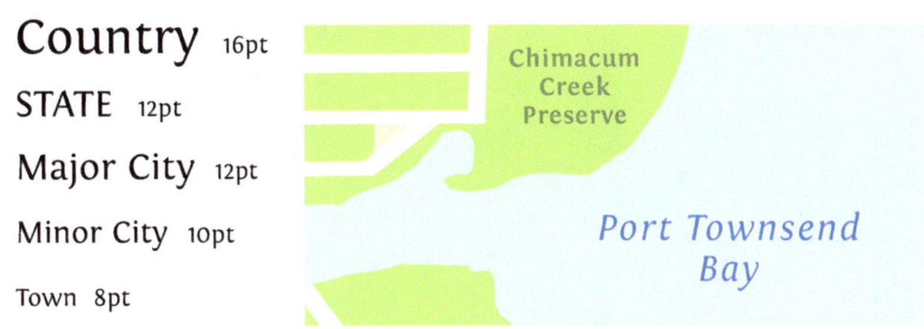

Aa	Bb	Cc	Dd	Ee	Ff	Gg	Hh
Ii	Jj	Kk	Ll	Mm	Nn	Oo	Pp
Qq	Rr	Ss	Tt	Uu	Vv	Ww	Xx
Yy	Zz	0	1	2	3	4	5
6	7	8	9				

All questions asked by five watched experts amaze the judge.
Fontin

All questions asked by five watched experts amaze the judge.
Fontin Bold

All questions asked by five watched experts amaze the judge.
Fontin Italic

ALL QUESTIONS ASKED BY FIVE WATCHED EXPERTS AMAZE THE JUDGE.
Fontin Small Caps

Fontin Sans designed by Jos Buivenga

Fontin Sans is an elegant, high-quality font with a slightly chiseled look. Use this for titles requiring classic personalities and features requiring no-fuss labels.

Country 16pt

STATE 12pt

Major City 12pt

Minor City 10pt

Town 8pt

E Arden St
E Arden St
6th Ave
5th Ave
E Moore St
E Moore St
Irondale
4th Ave
E Price St
E Price St

Aa	Bb	Cc	Dd	Ee	Ff	Gg	Hh
Ii	Jj	Kk	Ll	Mm	Nn	Oo	Pp
Qq	Rr	Ss	Tt	Uu	Vv	Ww	Xx
Yy	Zz	0	1	2	3	4	5
6	7	8	9				

All questions asked by five watched experts amaze the judge.
Fontin Sans

All questions asked by five watched experts amaze the judge.
Fontin Sans Bold

All questions asked by five watched experts amaze the judge.
Fontin Sans Italic

ALL QUESTIONS ASKED BY FIVE WATCHED EXPERTS AMAZE THE JUDGE.
Fontin Sans Small Caps

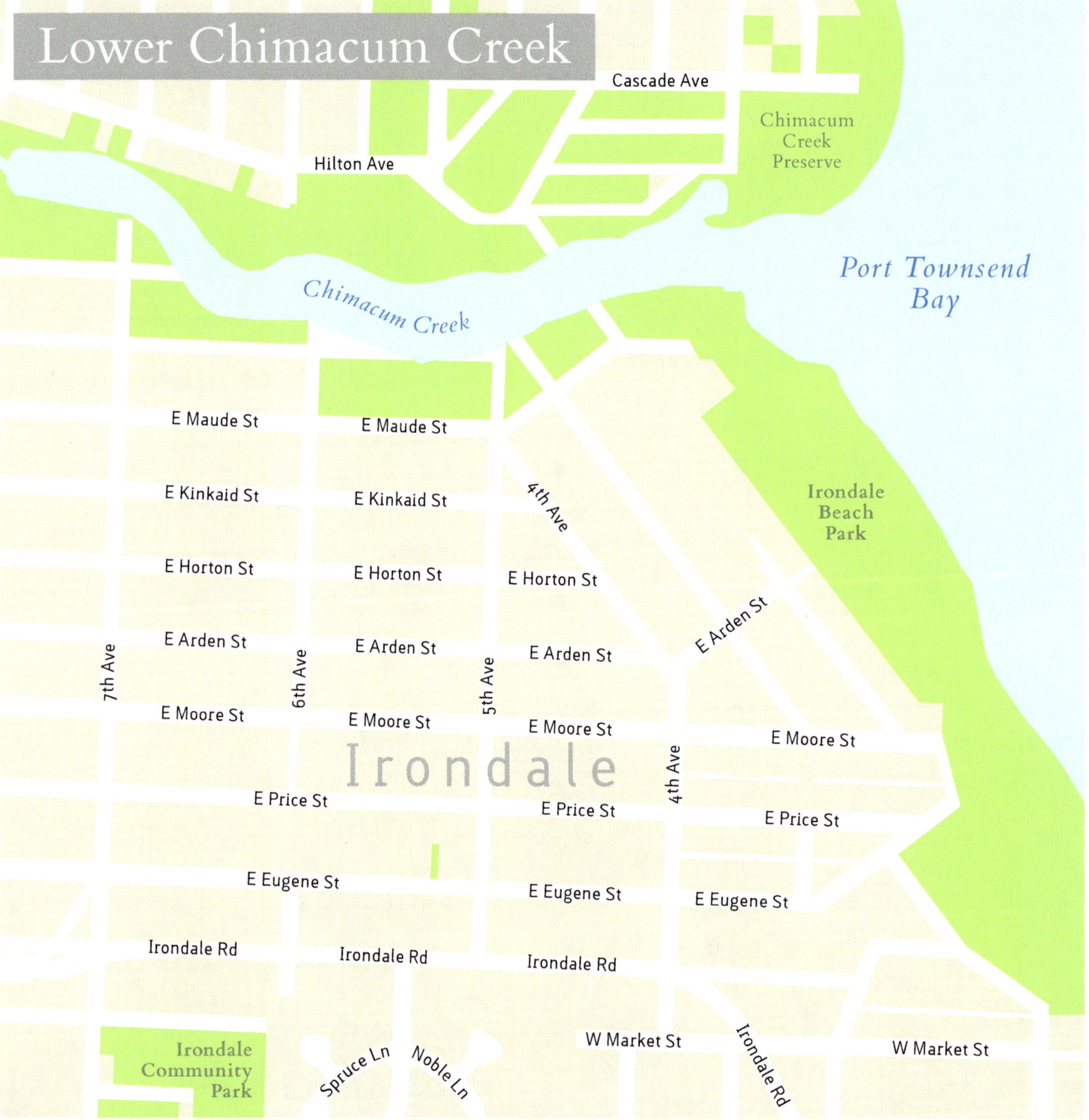
Lower Chimacum Creek
Cascade Ave
Hilton Ave
Chimacum Creek Preserve
Chimacum Creek
Port Townsend Bay
E Maude St
E Maude St
E Kinkaid St
E Kinkaid St
4th Ave
Irondale Beach Park
E Horton St
E Horton St
E Horton St
E Arden St
E Arden St
E Arden St
E Arden St
7th Ave
6th Ave
5th Ave
E Moore St
E Moore St
E Moore St
E Moore St
Irondale
4th Ave
E Price St
E Price St
E Price St
E Eugene St
E Eugene St
E Eugene St
Irondale Rd
Irondale Rd
Irondale Rd
W Market St
W Market St
Irondale Rd
Irondale Community Park
Spruce Ln
Noble Ln

Cardo designed by David Perry

Cardo is a traditional looking font that works well on antique style maps. Its italic format, a recent addition, is perfect for formal labels of natural features. A true bold was also recently added to the Cardo family.

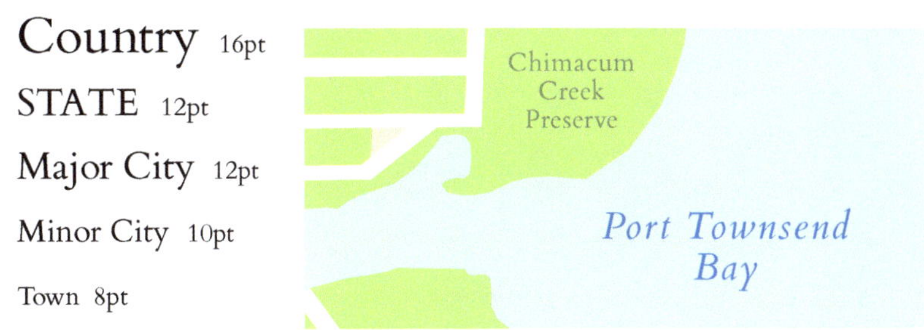

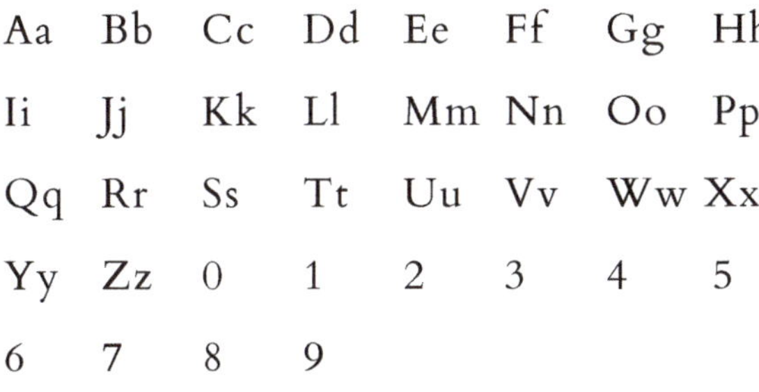

All questions asked by five watched experts amaze the judge.
Cardo

All questions asked by five watched experts amaze the judge.
Cardo Bold

All questions asked by five watched experts amaze the judge.
Cardo Italic

Lacuna designed by Peter Hoffmann

Lacuna has very tall x-heights for readability at small sizes. It has a regular and italic form but no true bold. In all, it possesses a certain quirkiness (e.g., slabs on i and f) but remains elegant.

Country 16pt
STATE 12pt
Major City 12pt
Minor City 10pt
Town 8pt

E Arden St
E Arden St
6th Ave
5th Ave
E Moore St
E Moore St
Irondale
4th Ave
E Price St
E Price St

Aa Bb Cc Dd Ee Ff Gg Hh
Ii Jj Kk Ll Mm Nn Oo Pp
Qq Rr Ss Tt Uu Vv Ww Xx
Yy Zz 0 1 2 3 4 5
6 7 8 9

All questions asked by five watched experts amaze the judge.
Lacuna Regular

All questions asked by five watched experts amaze the judge.
Lacuna Regular Italic

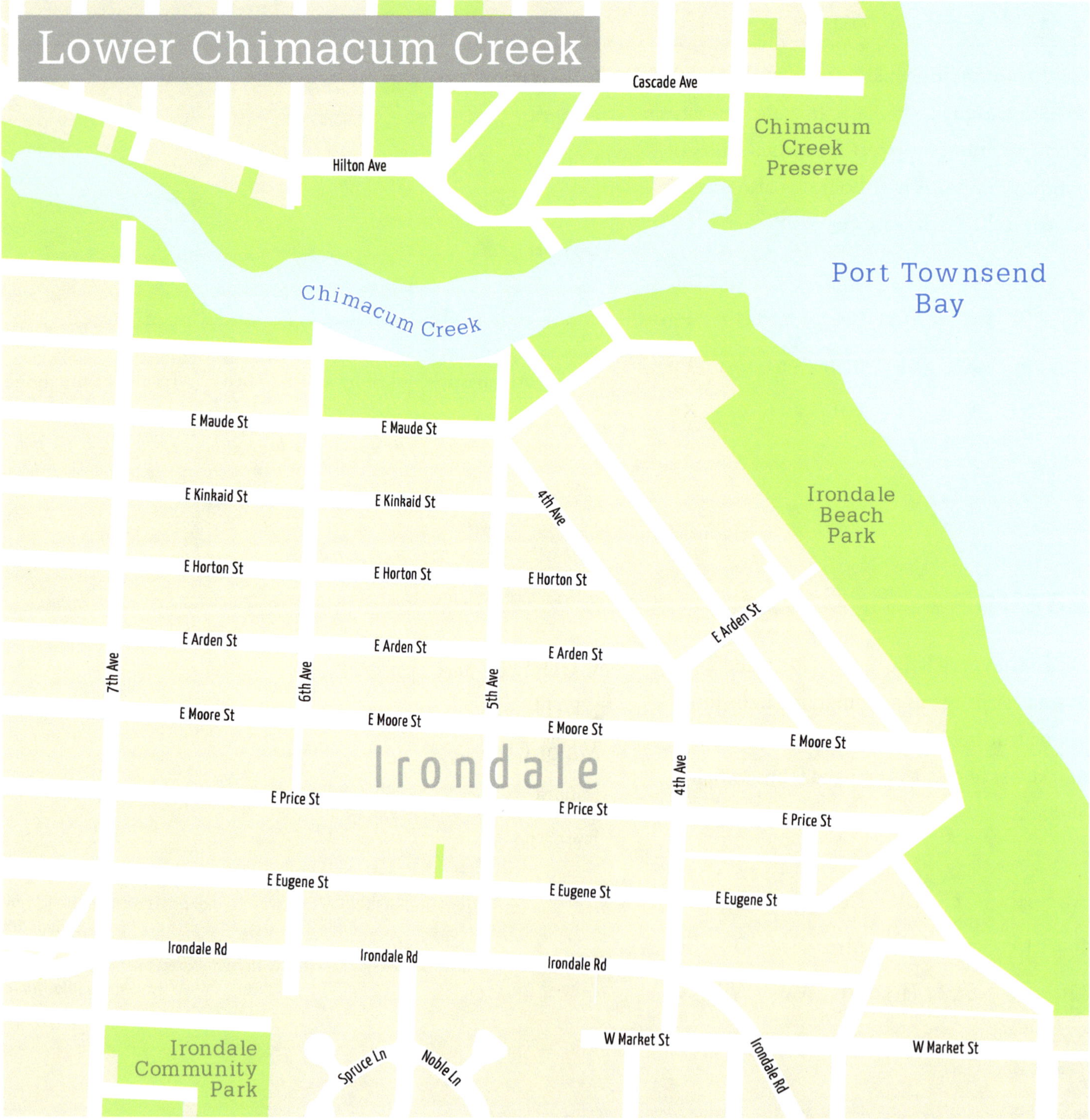

Lower Chimacum Creek

Cascade Ave
Chimacum Creek Preserve
Hilton Ave
Port Townsend Bay
Chimacum Creek
E Maude St
E Maude St
E Kinkaid St
E Kinkaid St
4th Ave
Irondale Beach Park
E Horton St
E Horton St
E Horton St
E Arden St
E Arden St
E Arden St
E Arden St
7th Ave
6th Ave
5th Ave
E Moore St
E Moore St
E Moore St
E Moore St
Irondale
4th Ave
E Price St
E Price St
E Price St
E Eugene St
E Eugene St
E Eugene St
Irondale Rd
Irondale Rd
Irondale Rd
W Market St
W Market St
Irondale Community Park
Spruce Ln
Noble Ln
Irondale Rd

Petit Latin designed by Manfred Klein

This funky and informal serif is very crisp and readable. It has a look that is somewhat reminiscent of a typewriter document. This font does not have true italics or true bold formats.

Country 16pt
STATE 12pt
Major City 12pt
Minor City 10pt
Town 8pt

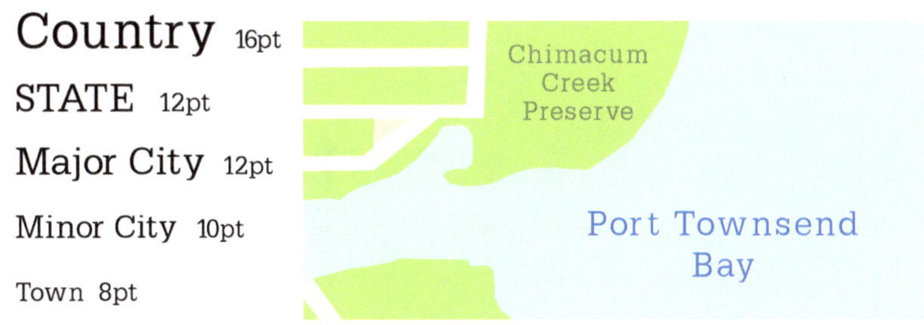

Aa Bb Cc Dd Ee Ff Gg Hh
Ii Jj Kk Ll Mm Nn Oo Pp
Qq Rr Ss Tt Uu Vv Ww Xx
Yy Zz 0 1 2 3 4 5
6 7 8 9

All questions asked by five watched experts amaze the judge.
Petit Latin

Yanone Kaffeesatz designed by Yanone

Yanone is a contemporary font with a firm weight and clean appeal for map labeling. While the bold version of this font was inspired by 1920s coffee house design, the other weights are more contemporary in personality. It is not very suitable for large blocks of text.

Country 16pt
STATE 12pt
Major City 12pt
Minor City 10pt
Town 8pt

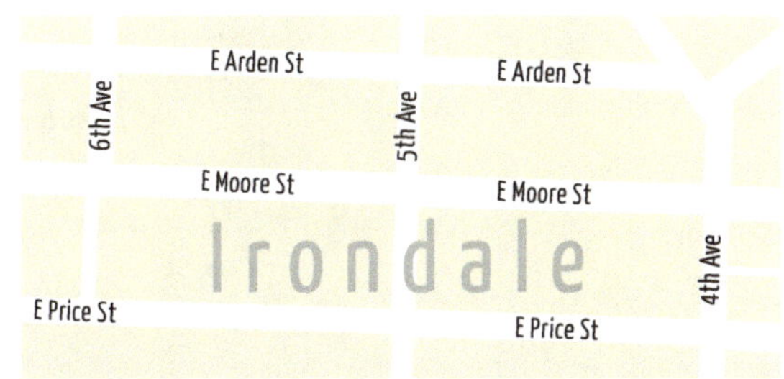

Aa Bb Cc Dd Ee Ff Gg Hh
Ii Jj Kk Ll Mm Nn Oo Pp
Qq Rr Ss Tt Uu Vv Ww Xx
Yy Zz 0 1 2 3 4 5
6 7 8 9

All questions asked by five watched experts amaze the judge.
Yanone Kaffeesatz Regular

All questions asked by five watched experts amaze the judge.
Yanone Kaffeesatz Light

All questions asked by five watched experts amaze the judge.
Yanone Kaffeesatz Bold

All questions asked by five watched experts amaze the judge.
Yanone Kaffeesatz Thin

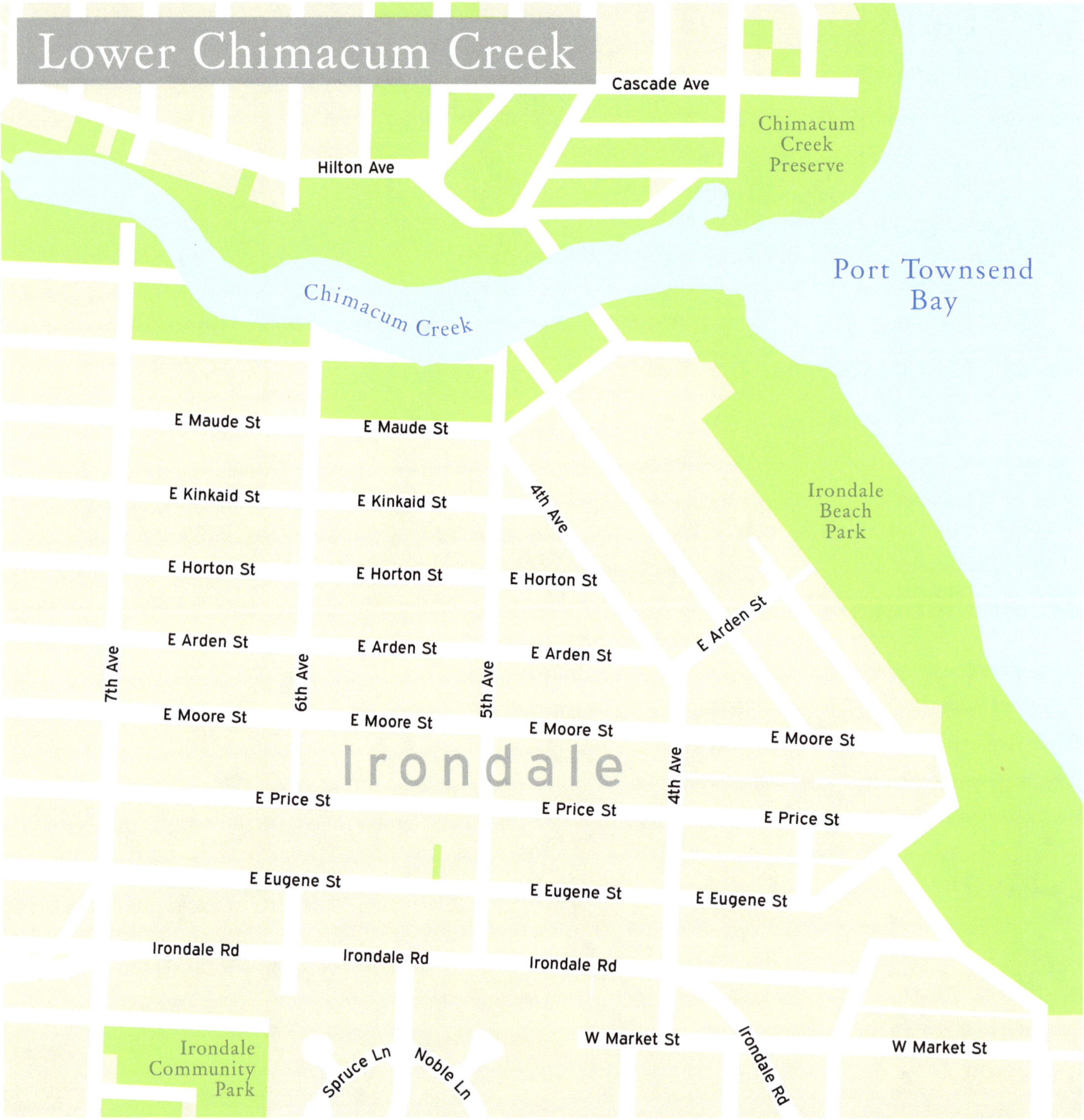

Lower Chimacum Creek

Cascade Ave
Chimacum Creek Preserve
Hilton Ave
Chimacum Creek
Port Townsend Bay
E Maude St
E Maude St
E Kinkaid St
E Kinkaid St
4th Ave
Irondale Beach Park
E Horton St
E Horton St
E Horton St
E Arden St
E Arden St
E Arden St
E Arden St
7th Ave
6th Ave
5th Ave
E Moore St
E Moore St
E Moore St
E Moore St
Irondale
4th Ave
E Price St
E Price St
E Price St
E Eugene St
E Eugene St
E Eugene St
Irondale Rd
Irondale Rd
Irondale Rd
W Market St
W Market St
Irondale Rd
Irondale Community Park
Spruce Ln
Noble Ln

Day Roman designed by Frederic Nader

Day Roman purposefully replicates some imperfections found in 16th century printing. It is not suitable for sizes greater than 30 pt. The 10-30 pt sizes are good for maps that require an old-style feel.

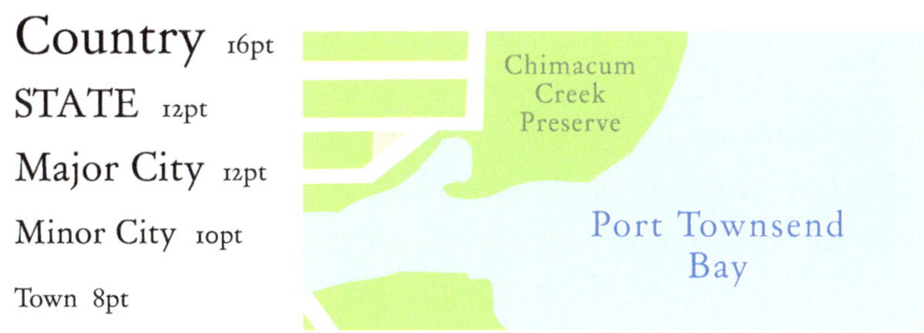

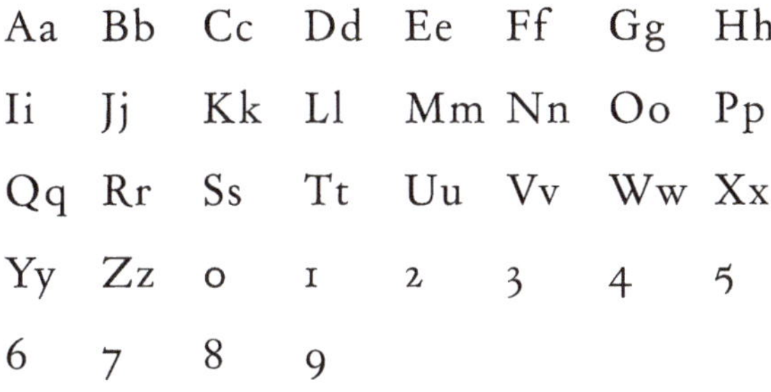

All questions asked by five watched experts amaze the judge.

Day Roman

Expressway designed by Ray Larabie

At large sizes, the letters in Expressway do not appear to sit on the same baseline, giving it a very informal feel. This font gives extra livelihood to titles and larger text blocks.

Country 16pt

STATE 12pt

Major City 12pt

Minor City 10pt

Town 8pt

E Arden St
6th Ave
5th Ave
E Arden St
E Moore St
E Moore St
Irondale
4th Ave
E Price St
E Price St

Aa Bb Cc Dd Ee Ff Gg Hh

Ii Jj Kk Ll Mm Nn Oo Pp

Qq Rr Ss Tt Uu Vv Ww Xx

Yy Zz 0 1 2 3 4 5

6 7 8 9

All questions asked by five watched experts amaze the judge.

Expressway Regular

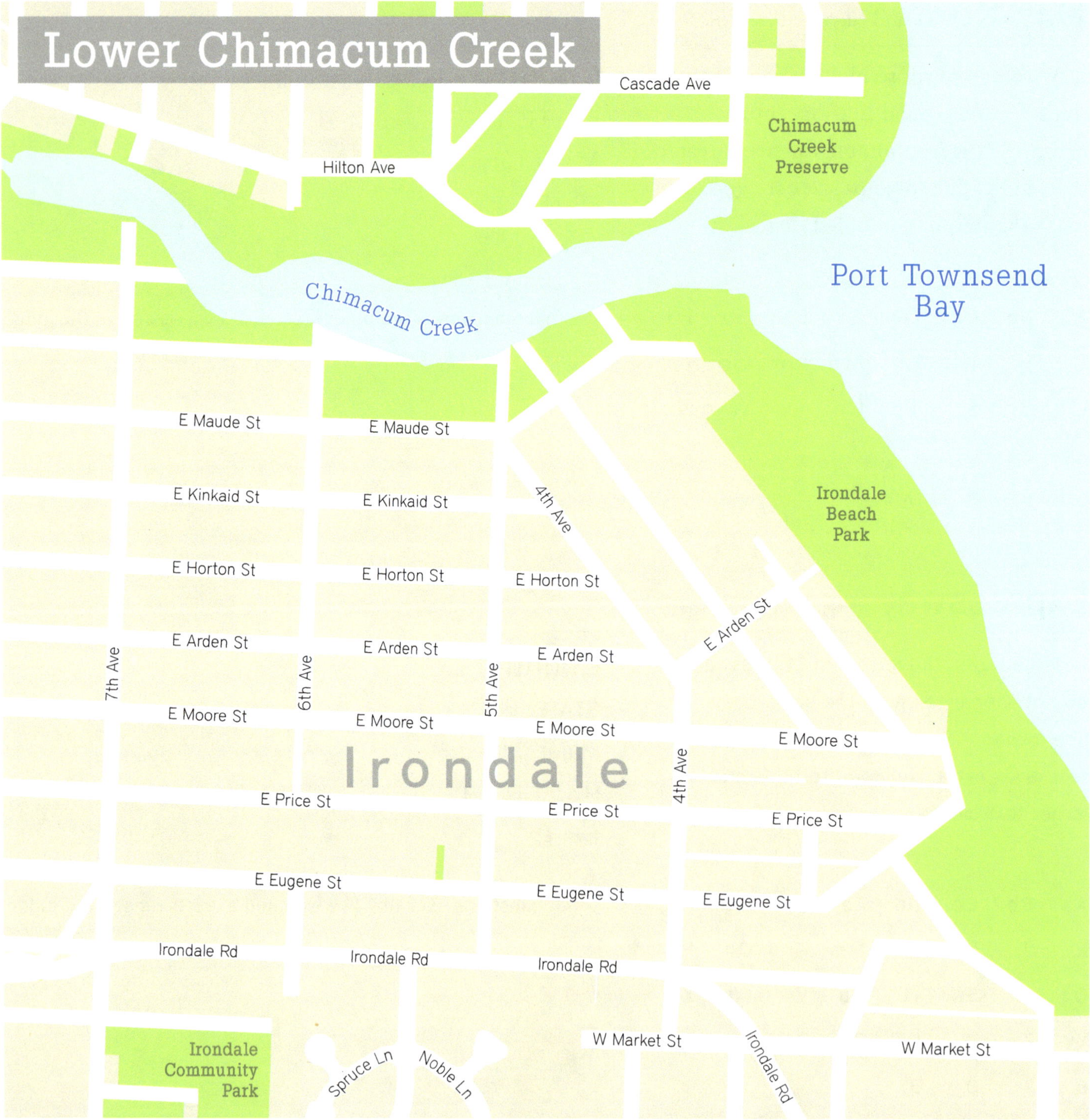

Lower Chimacum Creek

Cascade Ave
Chimacum Creek Preserve
Hilton Ave
Port Townsend Bay
Chimacum Creek
E Maude St
E Maude St
4th Ave
E Kinkaid St
E Kinkaid St
Irondale Beach Park
E Horton St
E Horton St
E Horton St
E Arden St
E Arden St
E Arden St
E Arden St
7th Ave
6th Ave
5th Ave
E Moore St
E Moore St
E Moore St
E Moore St
Irondale
4th Ave
E Price St
E Price St
E Price St
E Eugene St
E Eugene St
E Eugene St
Irondale Rd
Irondale Rd
Irondale Rd
W Market St
W Market St
Irondale Community Park
Spruce Ln
Noble Ln
Irondale Rd

Slab Tall X designed by Manfred Klein

As the name indicates, Slab Tall X is a slab serif with a tall x-height. It comes in two versions, the Regular (shown here) and Medium. The slabs lend extra weight to the tips of the glyphs, but not enough to overwhelm the text.

Country 16pt

STATE 12pt

Major City 12pt

Minor City 10pt

Town 8pt

Chimacum Creek Preserve

Port Townsend Bay

Aa	Bb	Cc	Dd	Ee	Ff	Gg	Hh
Ii	Jj	Kk	Ll	Mm	Nn	Oo	Pp
Qq	Rr	Ss	Tt	Uu	Vv	Ww	Xx
Yy	Zz	0	1	2	3	4	5
6	7	8	9				

All questions asked by five watched experts amaze the judge.
Slab Tall X

All questions asked by five watched experts amaze the judge.
Slab Tall X Medium

District designed by Dylan Smith and Kienan Smith

District Thin (shown here) is available for free but the 13 other varieties of this typeface must be purchased. Its clean, no-nonsense personality and tall x-height make it suitable for small labels.

Country 16pt

STATE 12pt

Major City 12pt

Minor City 10pt

Town 8pt

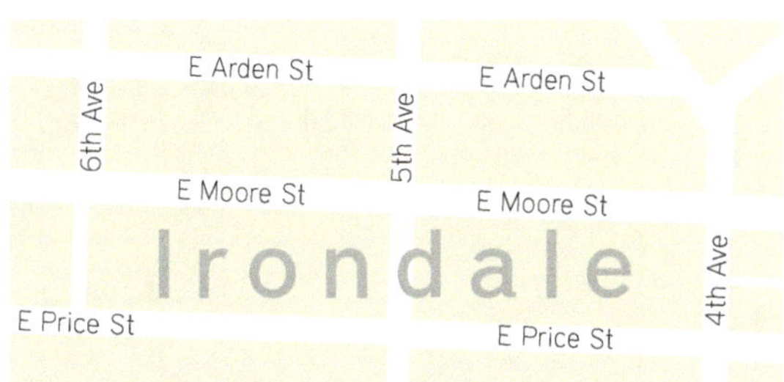

Aa	Bb	Cc	Dd	Ee	Ff	Gg	Hh
Ii	Jj	Kk	Ll	Mm	Nn	Oo	Pp
Qq	Rr	Ss	Tt	Uu	Vv	Ww	Xx
Yy	Zz	0	1	2	3	4	5
6	7	8	9				

All questions asked by five watched experts amaze the judge.
District Thin

All questions asked by five watched experts amaze the judge.
District Medium

All questions asked by five watched experts amaze the judge.
District Thin Italic

All questions asked by five watched experts amaze the judge.
District Medium Italic

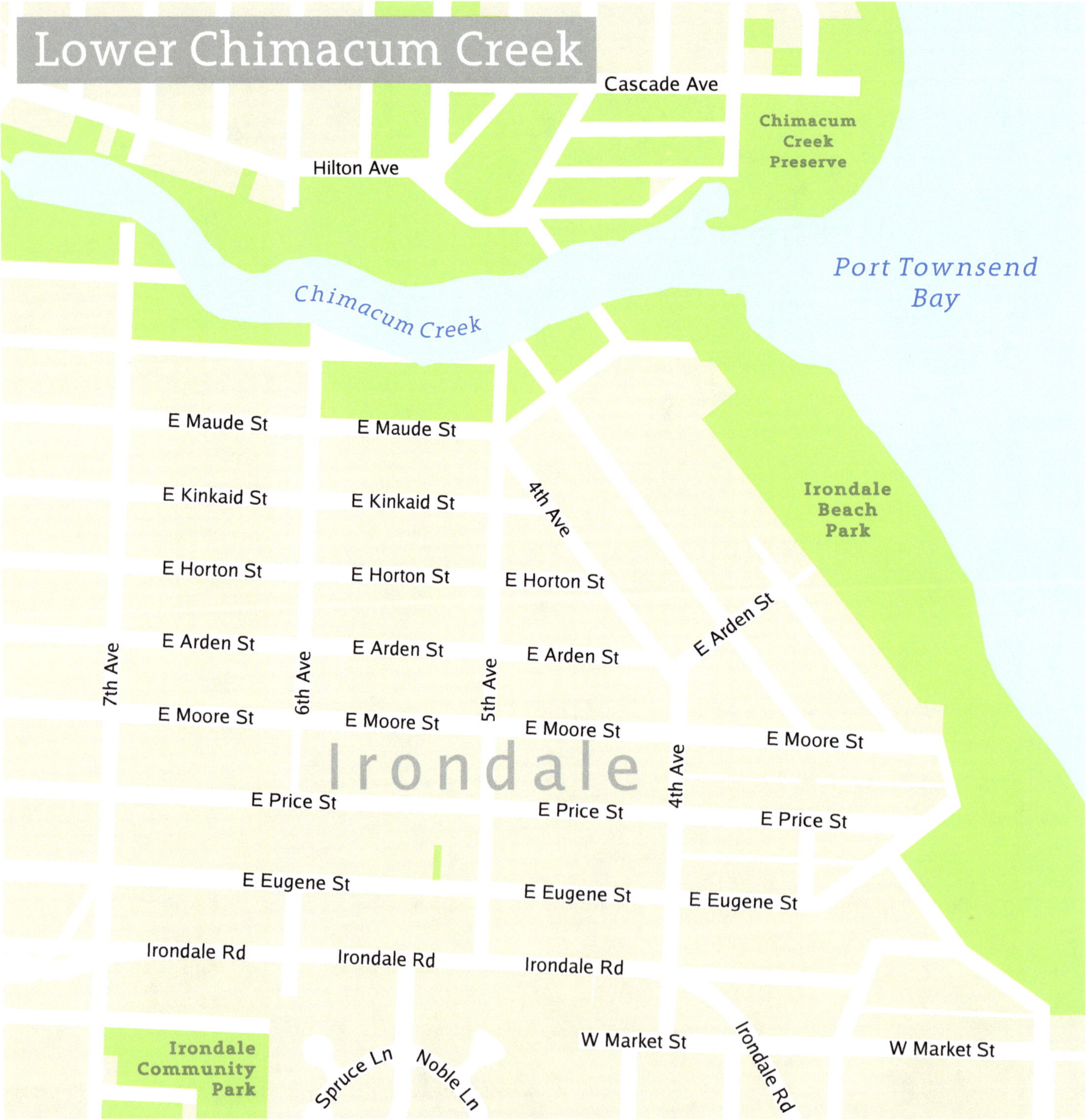

Lower Chimacum Creek

Cascade Ave
Chimacum Creek Preserve
Hilton Ave
Port Townsend Bay
Chimacum Creek
E Maude St
E Maude St
E Kinkaid St
E Kinkaid St
4th Ave
Irondale Beach Park
E Horton St
E Horton St
E Horton St
E Arden St
E Arden St
E Arden St
E Arden St
7th Ave
6th Ave
5th Ave
E Moore St
E Moore St
E Moore St
E Moore St
Irondale
4th Ave
E Price St
E Price St
E Price St
E Eugene St
E Eugene St
E Eugene St
Irondale Rd
Irondale Rd
Irondale Rd
W Market St
Irondale Rd
W Market St
Irondale Community Park
Spruce Ln
Noble Ln

Grandesign Neue designed by Unknown

Grandesign Neue has several varieties including a sans serif. The regular serif version is a fairly straightforward design with slab-like serifs. The bold version has more personality for interesting titles.

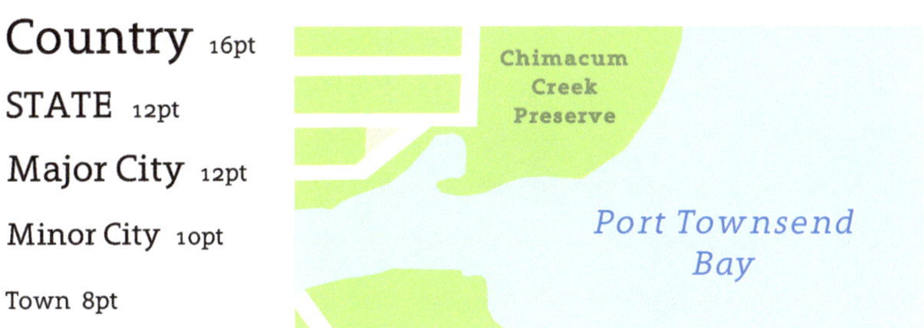

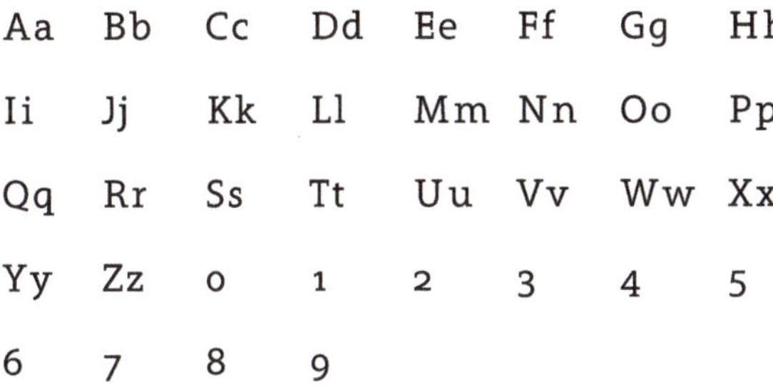

All questions asked by five watched experts amaze the judge.
Grandesign Neue Serif

All questions asked by five watched experts amaze the judge.
Grandesign Neue Serif Bold

All questions asked by five watched experts amaze the judge.
Grandesign Neue Serif Italic

All questions asked by five watched experts amaze the judge.
Grandesign Neue Serif Bold Italic

Luxi Sans designed by Kris Holmes and Charles Bigelow

Luxi Sans can also be paired with its relatives Luxi Serif and Luxi Mono. With extreme variance in line widths, the letters in this typeface have an informal feel at large text sizes.

Country 16pt
STATE 12pt
Major City 12pt
Minor City 10pt
Town 8pt

E Arden St
E Arden St
6th Ave
5th Ave
E Moore St
E Moore St
Irondale
4th Ave
E Price St
E Price St

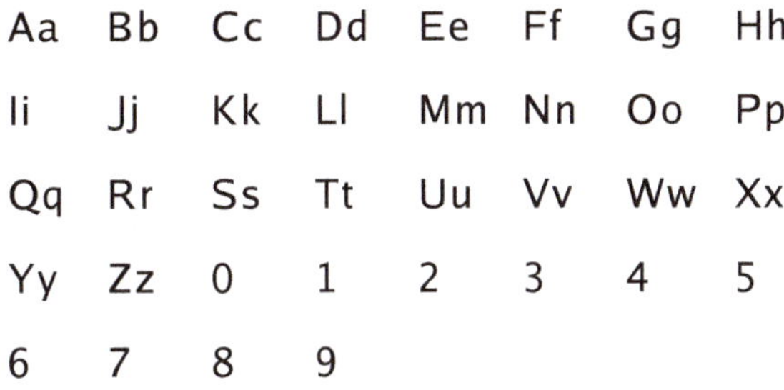

All questions asked by five watched experts amaze the judge.
Luxi Sans

All questions asked by five watched experts amaze the judge.
Luxi Sans Bold

All questions asked by five watched experts amaze the judge.
Luxi Sans Oblique

All questions asked by five watched experts amaze the judge.
Luxi Sans Bold Oblique

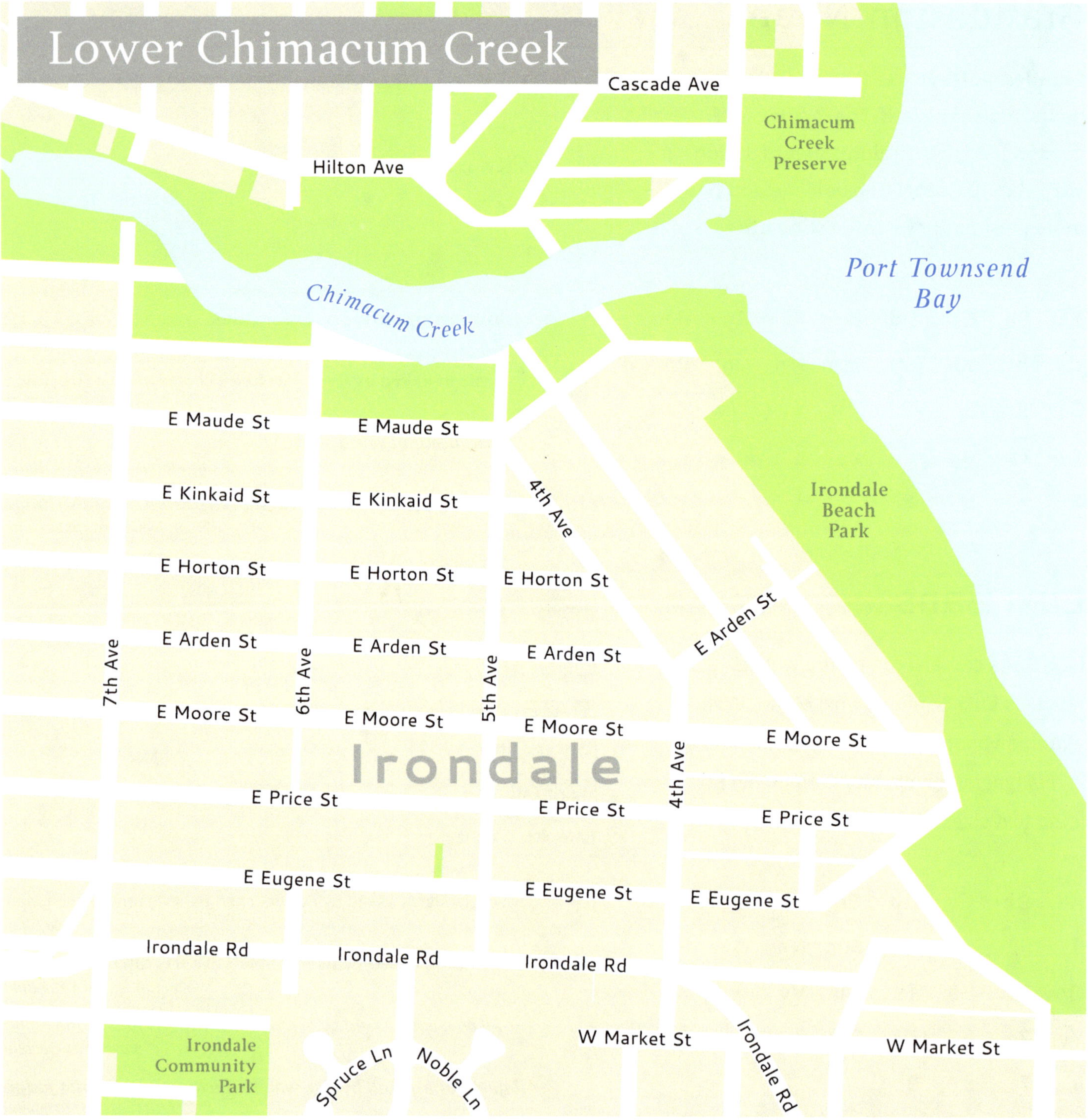

Lower Chimacum Creek

Cascade Ave
Hilton Ave
Chimacum Creek Preserve
Port Townsend Bay
Chimacum Creek
E Maude St
E Maude St
E Kinkaid St
E Kinkaid St
4th Ave
Irondale Beach Park
E Horton St
E Horton St
E Horton St
E Arden St
E Arden St
E Arden St
E Arden St
7th Ave
6th Ave
5th Ave
E Moore St
E Moore St
E Moore St
E Moore St
4th Ave
Irondale
E Price St
E Price St
E Price St
E Eugene St
E Eugene St
E Eugene St
Irondale Rd
Irondale Rd
Irondale Rd
W Market St
W Market St
Irondale Rd
Irondale Community Park
Spruce Ln
Noble Ln

Ingleby designed by David Engelby

Ingleby works well on traditional or antique style maps. It has four varieties, all with fine styling and weight. While it is well crafted, the x-heights do vary—especially in the italic version. Some characters show calligraphy-like end strokes.

Country 16pt

STATE 12pt

Major City 12pt

Minor City 10pt

Town 8pt

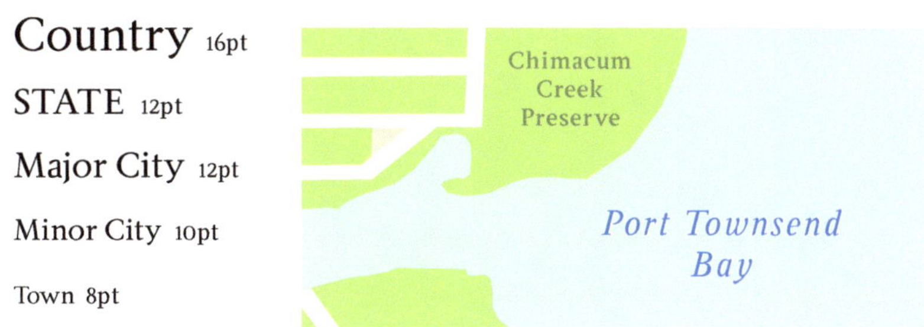

Aa Bb Cc Dd Ee Ff Gg Hh

Ii Jj Kk Ll Mm Nn Oo Pp

Qq Rr Ss Tt Uu Vv Ww Xx

Yy Zz 0 1 2 3 4 5

6 7 8 9

All questions asked by five watched experts amaze the judge.
Ingleby Regular

All questions asked by five watched experts amaze the judge.
Ingleby Bold

All questions asked by five watched experts amaze the judge.
InglebyItalic

All questions asked by five watched experts amaze the judge.
Ingleby Bold Italic

Cantarell designed by Dave Crossland

Cantarell is optimized for reading at small sizes on smaller digital devices. The print output of the bold and italic version may not produce desired results. This font has a conventional feel.

Country 16pt

STATE 12pt

Major City 12pt

Minor City 10pt

Town 8pt

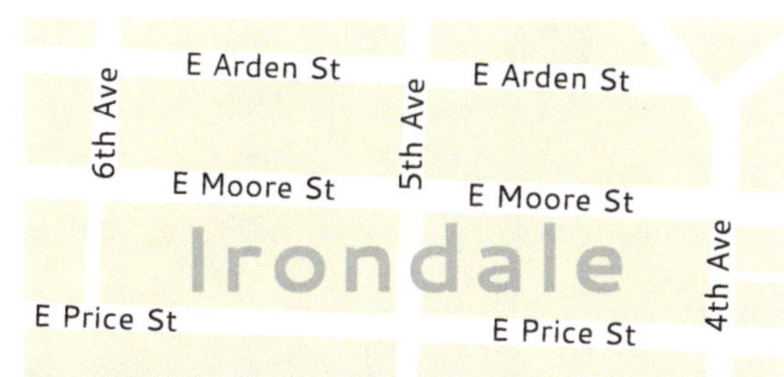

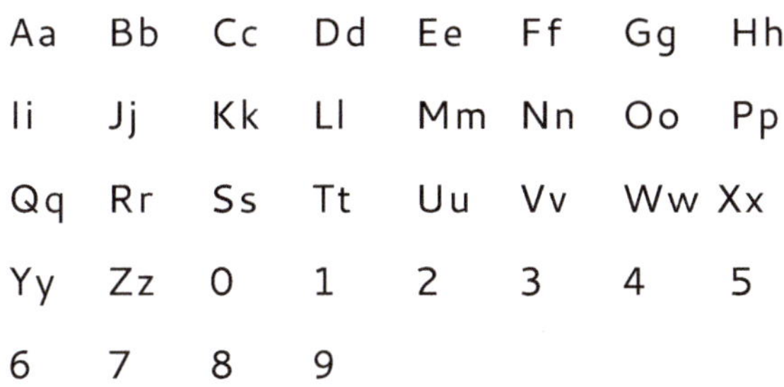

All questions asked by five watched experts amaze the judge.
Cantarell

All questions asked by five watched experts amaze the judge.
Cantarell Bold

All questions asked by five watched experts amaze the judge.
Cantarell Oblique

All questions asked by five watched experts amaze the judge.
Cantarell Bold Oblique

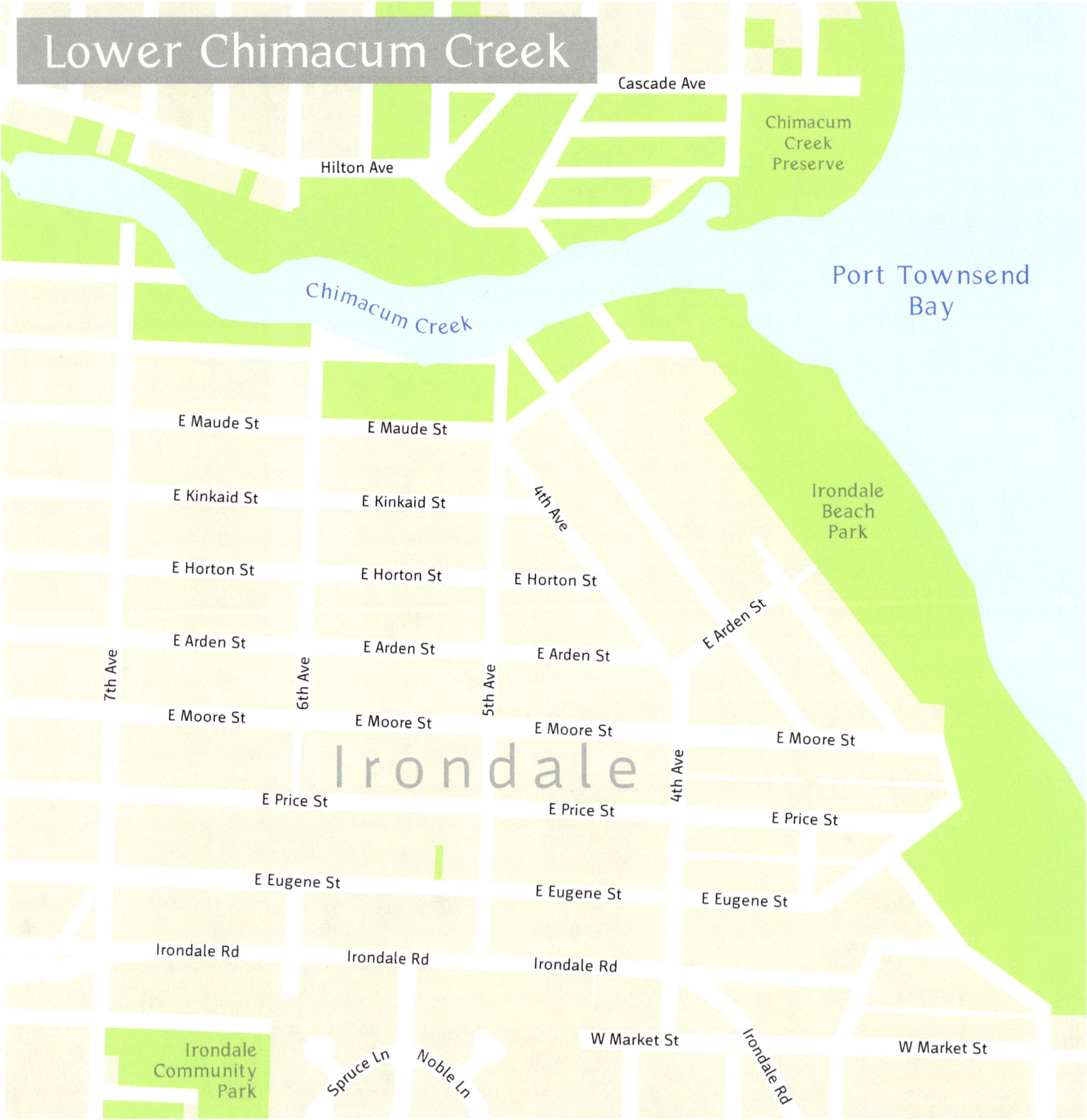

Lower Chimacum Creek

Cascade Ave
Hilton Ave
Chimacum Creek Preserve
Port Townsend Bay
Chimacum Creek
E Maude St
E Maude St
E Kinkaid St
E Kinkaid St
4th Ave
Irondale Beach Park
E Horton St
E Horton St
E Horton St
E Arden St
E Arden St
E Arden St
E Arden St
E Arden St
7th Ave
6th Ave
5th Ave
E Moore St
E Moore St
E Moore St
E Moore St
Irondale
4th Ave
E Price St
E Price St
E Price St
E Eugene St
E Eugene St
E Eugene St
Irondale Rd
Irondale Rd
Irondale Rd
W Market St
W Market St
Irondale Community Park
Spruce Ln
Noble Ln
Irondale Rd

Magnificent designed by Diego Vallejo

Magnificent is a very informal serif design reminiscent of handwriting. While it only has one weight and style it may be useful in some contexts as it can sometimes be difficult to find a serif with an informal character.

Country 16pt

STATE 12pt

Major City 12pt

Minor City 10pt

Town 8pt

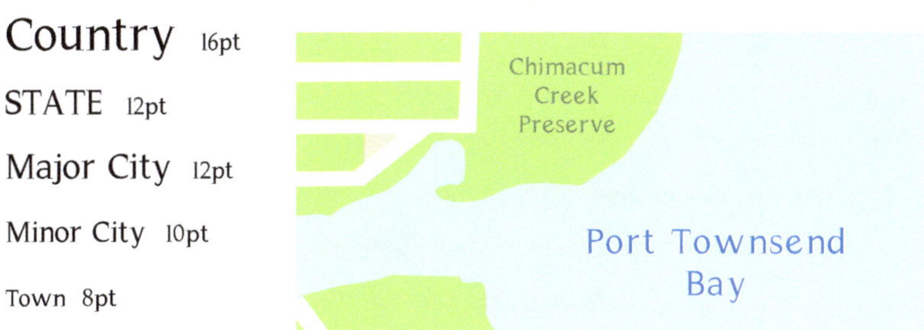

Aa Bb Cc Dd Ee Ff Gg Hh

Ii Jj Kk Ll Mm Nn Oo Pp

Qq Rr Ss Tt Uu Vv Ww Xx

Yy Zz 0 1 2 3 4 5

6 7 8 9

All questions asked by five watched experts amaze the judge.
Magnificent

Mr Jones designed by Richard Miller

The book and book italic versions of Mr Jones are free. This font has wide lowercase letters for legibility at small sizes and a consistent appearance. Tracking must be decreased at large sizes.

Country 16pt

STATE 12pt

Major City 12pt

Minor City 10pt

Town 8pt

E Arden St
E Arden St
6th Ave
5th Ave
E Moore St
E Moore St
Irondale
4th Ave
E Price St
E Price St

Aa Bb Cc Dd Ee Ff Gg Hh

Ii Jj Kk Ll Mm Nn Oo Pp

Qq Rr Ss Tt Uu Vv Ww Xx

Yy Zz 0 1 2 3 4 5

6 7 8 9

All questions asked by five watched experts amaze the judge.
Mr Jones Book

All questions asked by five watched experts amaze the judge.
Mr Jones Book Italic

All questions asked by five watched experts amaze the judge.
Mr Jones Bold

All questions asked by five watched experts amaze the judge.
Mr Jones Bold Italic

FOR FEE TYPEFACES

These pages showcase 10 typefaces that can be purchased and that are particularly well suited for maps. Most cartographers buy at least a few of these during their career and some buy many. These typefaces often have a wider range of weights and styles, including condensed and true italic forms that are invaluable to the professional cartographer.

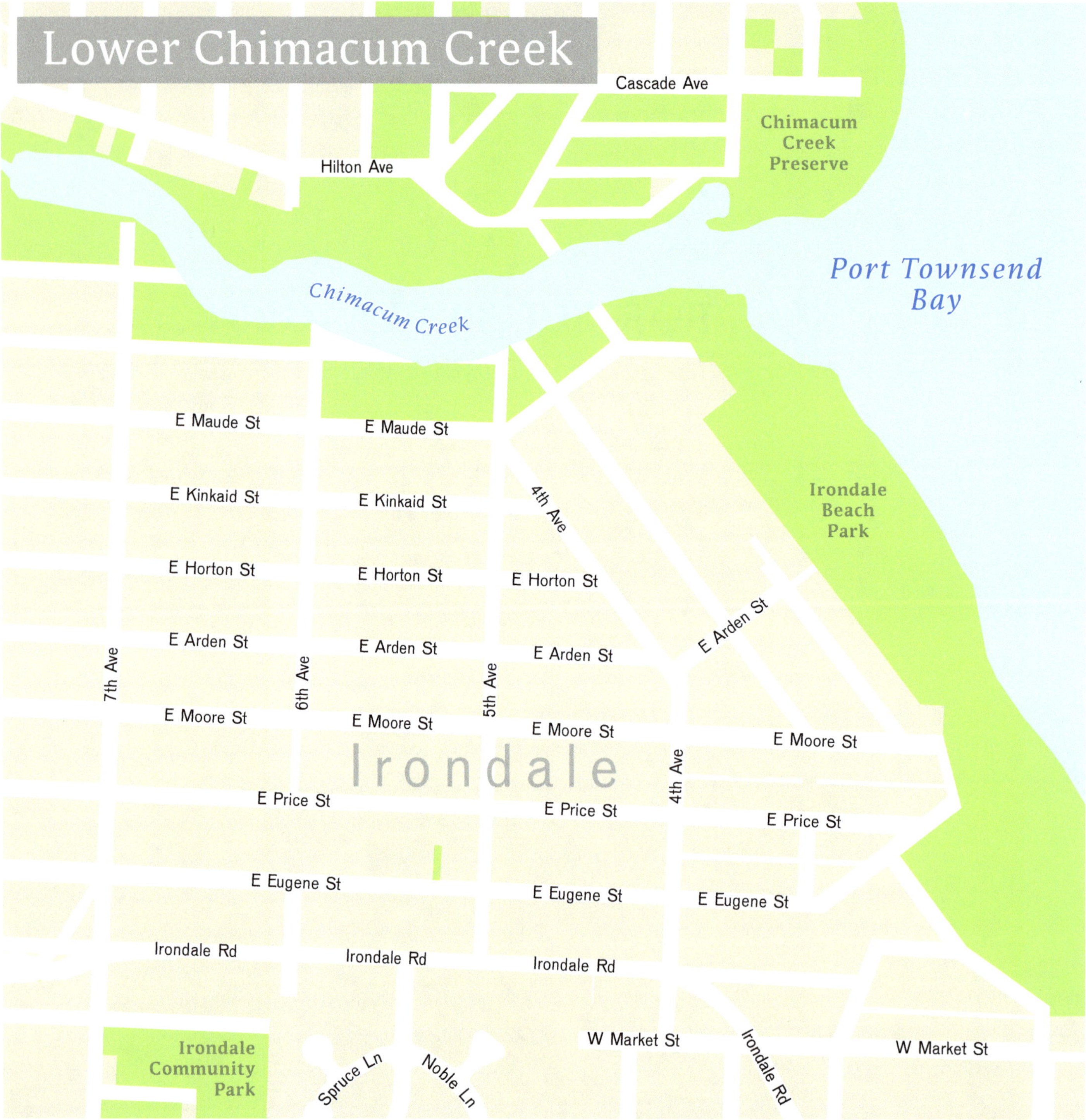

Lower Chimacum Creek

Cascade Ave
Chimacum Creek Preserve
Hilton Ave
Port Townsend Bay
Chimacum Creek
E Maude St
E Maude St
E Kinkaid St
E Kinkaid St
4th Ave
Irondale Beach Park
E Horton St
E Horton St
E Horton St
E Arden St
E Arden St
E Arden St
E Arden St
7th Ave
6th Ave
5th Ave
E Moore St
E Moore St
E Moore St
E Moore St
Irondale
4th Ave
E Price St
E Price St
E Price St
E Eugene St
E Eugene St
E Eugene St
Irondale Rd
Irondale Rd
Irondale Rd
W Market St
W Market St
Irondale Rd
Irondale Community Park
Spruce Ln
Noble Ln

FF Parable designed by Christopher Burke

FF Parable looks best between 6 and 10 pt text sizes. With its short descenders and distinctive italics, it is a great choice for stream labeling. It is also superb for maps with many levels of hierarchy.

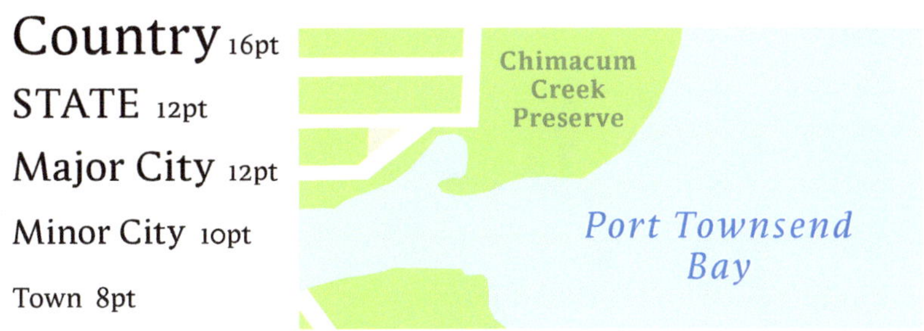

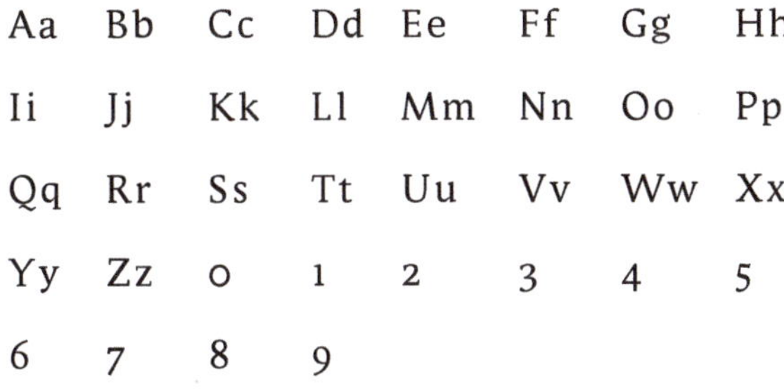

All questions asked by five watched experts amaze the judge.
FF Parable

All questions asked by five watched experts amaze the judge.
FF Parable Bold

All questions asked by five watched experts amaze the judge.
FF Parable Italic

All questions asked by five watched experts amaze the judge.
FF Parable Bold Italic

Venus Mager designed by Bauersche Giesserei

Venus Mager is a compact sans serif with extra height. It has a modern appeal with its unadorned and straightforward personality. The Mager variety is suitable for feature labels.

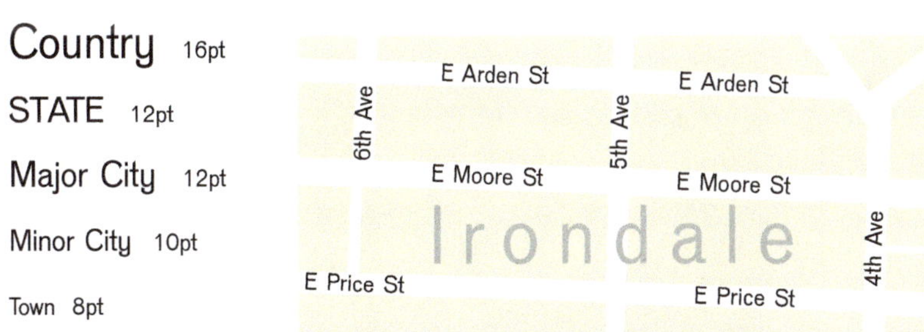

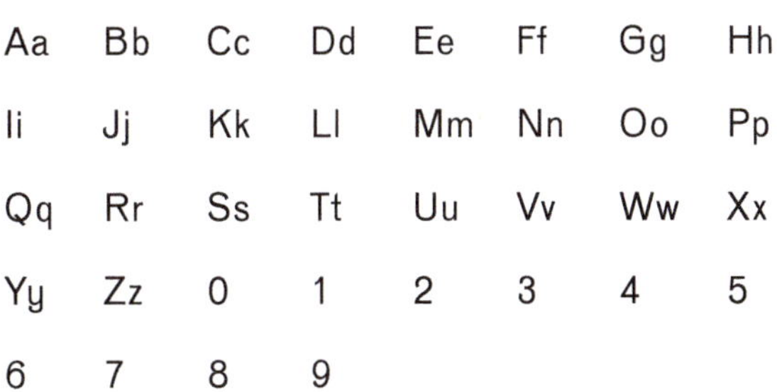

All questions asked by five watched experts amaze the judge.
Venus Mager

All questions asked by five watched experts amaze the judge.
Venus Mager Light Italic

All questions asked by five watched experts amaze the judge.
Venus Mager Reclining

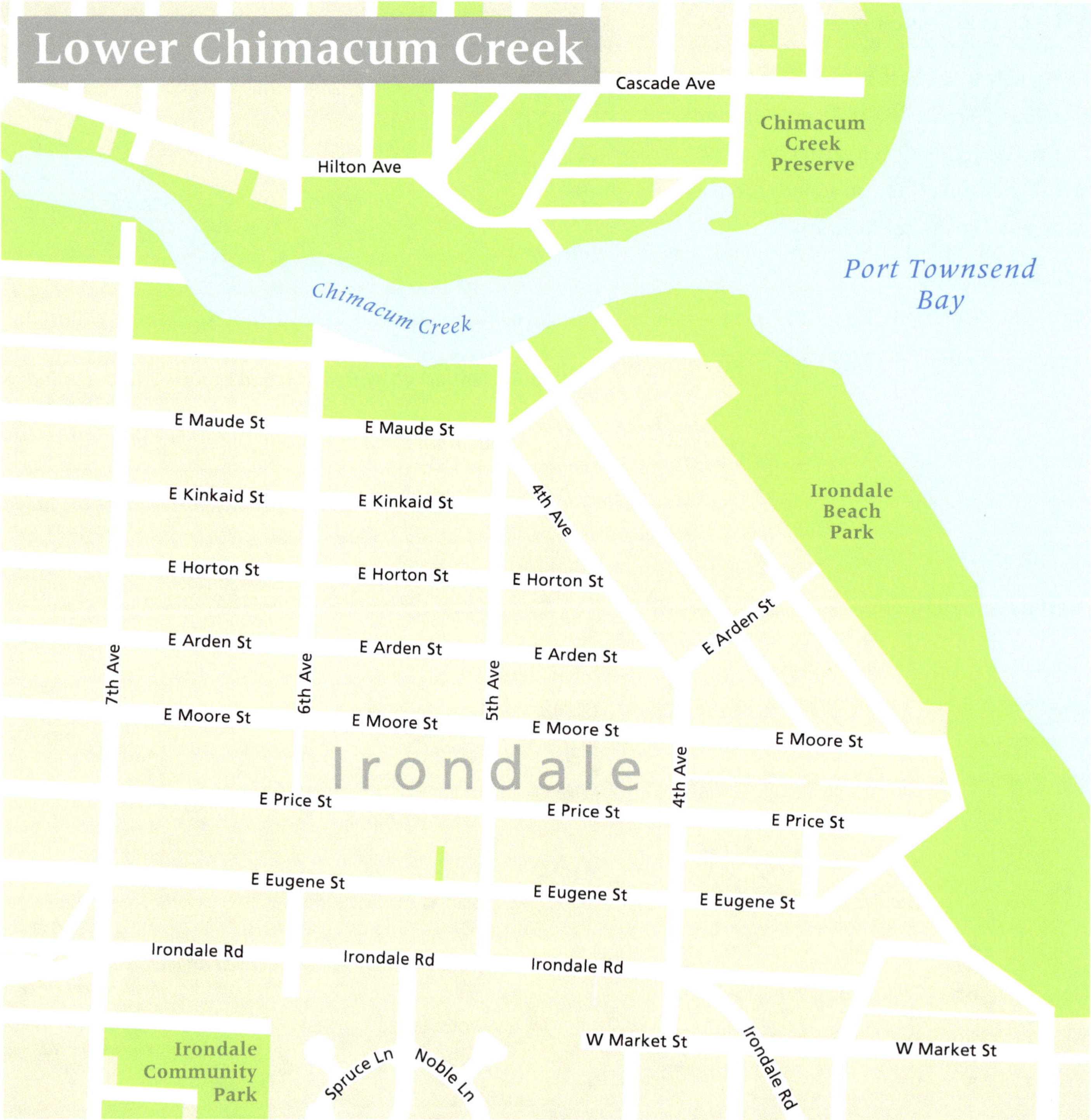

Lower Chimacum Creek

Cascade Ave
Chimacum Creek Preserve
Hilton Ave
Port Townsend Bay
Chimacum Creek
E Maude St
E Maude St
E Kinkaid St
E Kinkaid St
4th Ave
Irondale Beach Park
E Horton St
E Horton St
E Horton St
E Arden St
E Arden St
E Arden St
E Arden St
7th Ave
6th Ave
5th Ave
E Moore St
E Moore St
E Moore St
E Moore St
Irondale
4th Ave
E Price St
E Price St
E Price St
E Eugene St
E Eugene St
E Eugene St
Irondale Rd
Irondale Rd
Irondale Rd
W Market St
Irondale Rd
W Market St
Irondale Community Park
Spruce Ln
Noble Ln

Frutiger Serif designed by Adrian Frutiger

Frutiger Serif was released in 2008 as a serif companion to the popular sans version of the typeface. While the sans version is relatively simple looking, the serif version has an air of elegance to it.

Country 16pt
STATE 12pt
Major City 12pt
Minor City 10pt
Town 8pt

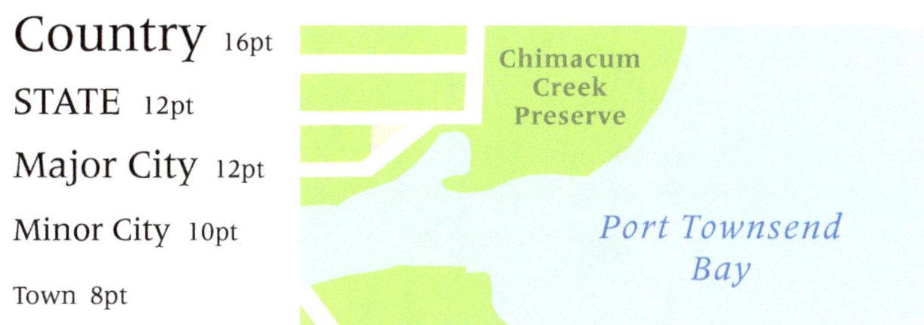

Aa Bb Cc Dd Ee Ff Gg Hh
Ii Jj Kk Ll Mm Nn Oo Pp
Qq Rr Ss Tt Uu Vv Ww Xx
Yy Zz 0 1 2 3 4 5
6 7 8 9

All questions asked by five watched experts amaze the judge.
Frutiger Serif

All questions asked by five watched experts amaze the judge.
Frutiger Serif Bold

All questions asked by five watched experts amaze the judge.
Frutiger Serif Italic

All questions asked by five watched experts amaze the judge.
Frutiger Serif Condensed

Frutiger designed by Adrian Frutiger

Frutiger is the original sans serif version of the Frutiger typeface. It is a simple looking font with many options for the professional cartographer to create complex hierarchies.

Country 16pt
STATE 12pt
Major City 12pt
Minor City 10pt
Town 8pt

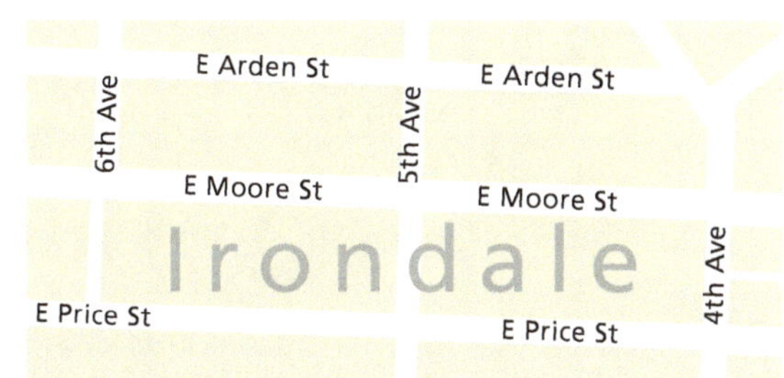

Aa Bb Cc Dd Ee Ff Gg Hh
Ii Jj Kk Ll Mm Nn Oo Pp
Qq Rr Ss Tt Uu Vv Ww Xx
Yy Zz 0 1 2 3 4 5
6 7 8 9

All questions asked by five watched experts amaze the judge.
Frutiger

All questions asked by five watched experts amaze the judge.
Frutiger Bold

All questions asked by five watched experts amaze the judge.
Frutiger Italic

All questions asked by five watched experts amaze the judge.
Frutiger Condensed

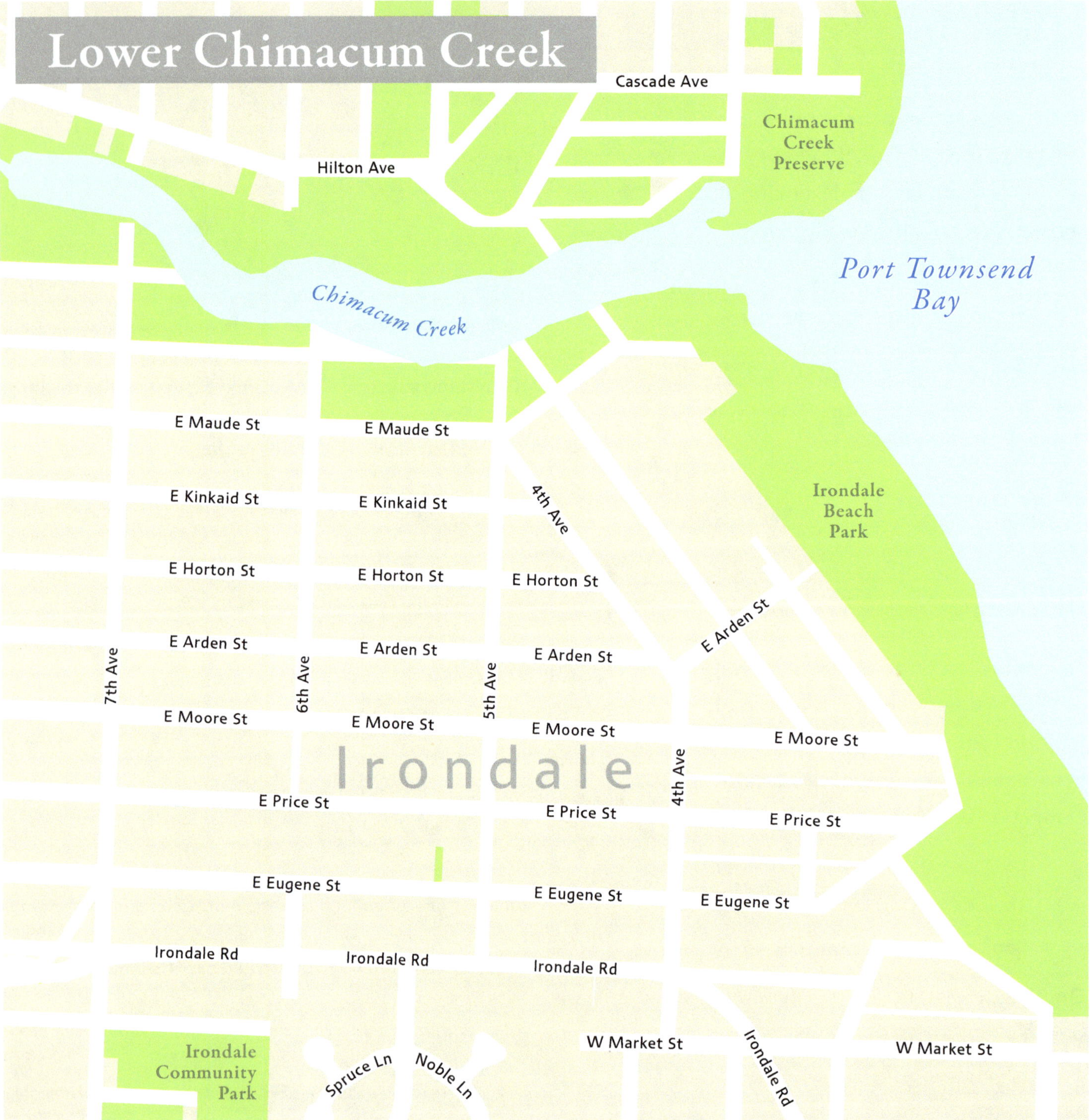
Lower Chimacum Creek
Cascade Ave
Chimacum Creek Preserve
Hilton Ave
Port Townsend Bay
Chimacum Creek
E Maude St
E Maude St
E Kinkaid St
E Kinkaid St
4th Ave
Irondale Beach Park
E Horton St
E Horton St
E Horton St
E Arden St
E Arden St
E Arden St
E Arden St
E Arden St
7th Ave
6th Ave
5th Ave
E Moore St
E Moore St
E Moore St
E Moore St
Irondale
4th Ave
E Price St
E Price St
E Price St
E Eugene St
E Eugene St
E Eugene St
Irondale Rd
Irondale Rd
Irondale Rd
W Market St
W Market St
Irondale Rd
Irondale Community Park
Spruce Ln
Noble Ln

Garamond Premier Pro designed by Robert Slimbach

Garamond Premier Pro is a widely used typeface that has elements of elegance and sophistication without undue flourish. It is a great choice for situations where a sense of visual seriousness is especially needed.

Country 16pt

STATE 12pt

Major City 12pt

Minor City 10pt

Town 8pt

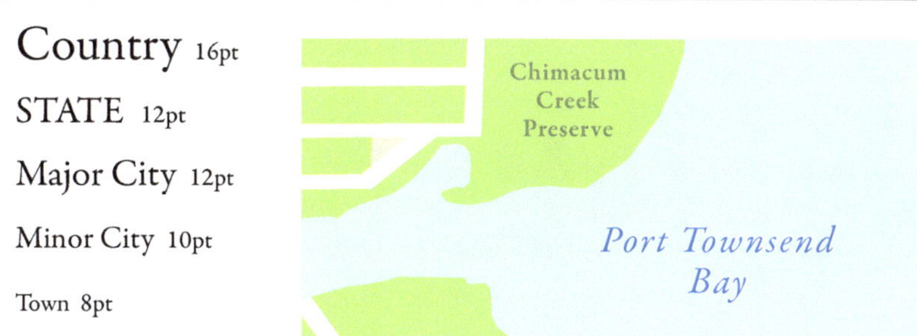

Aa Bb Cc Dd Ee Ff Gg Hh

Ii Jj Kk Ll Mm Nn Oo Pp

Qq Rr Ss Tt Uu Vv Ww Xx

Yy Zz 0 1 2 3 4 5

6 7 8 9

All questions asked by five watched experts amaze the judge.
Garamond Premier Pro

All questions asked by five watched experts amaze the judge.
Garamond Premier Pro Semibold

All questions asked by five watched experts amaze the judge.
Garamond Premier Pro Italic

All questions asked by five watched experts amaze the judge.
Garamond Premier Pro Semibold Italic

Cisalpin designed by Felix Arnold

Designed specifically for cartography, Cisalpin is very readable at small sizes and when layered on detailed background images. It has straight, narrow lines and steep, angled italics.

Country 16pt

STATE 12pt

Major City 12pt

Minor City 10pt

Town 8pt

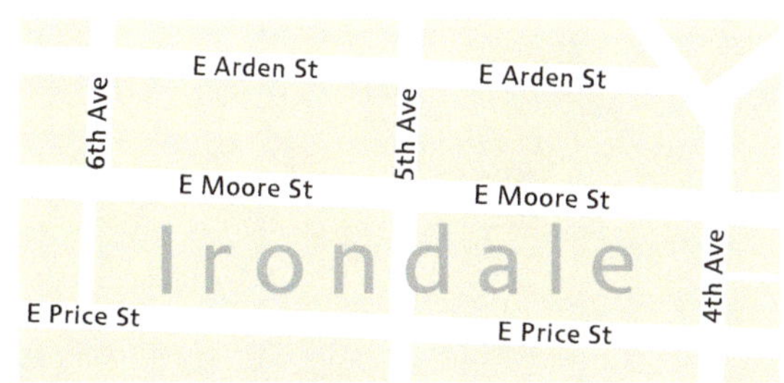

Aa Bb Cc Dd Ee Ff Gg Hh

Ii Jj Kk Ll Mm Nn Oo Pp

Qq Rr Ss Tt Uu Vv Ww Xx

Yy Zz 0 1 2 3 4 5

6 7 8 9

All questions asked by five watched experts amaze the judge.
Cisalpin

All questions asked by five watched experts amaze the judge.
Cisalpin Bold

All questions asked by five watched experts amaze the judge.
Cisalpin Italic

All questions asked by five watched experts amaze the judge.
Cisalpin Bold Italic

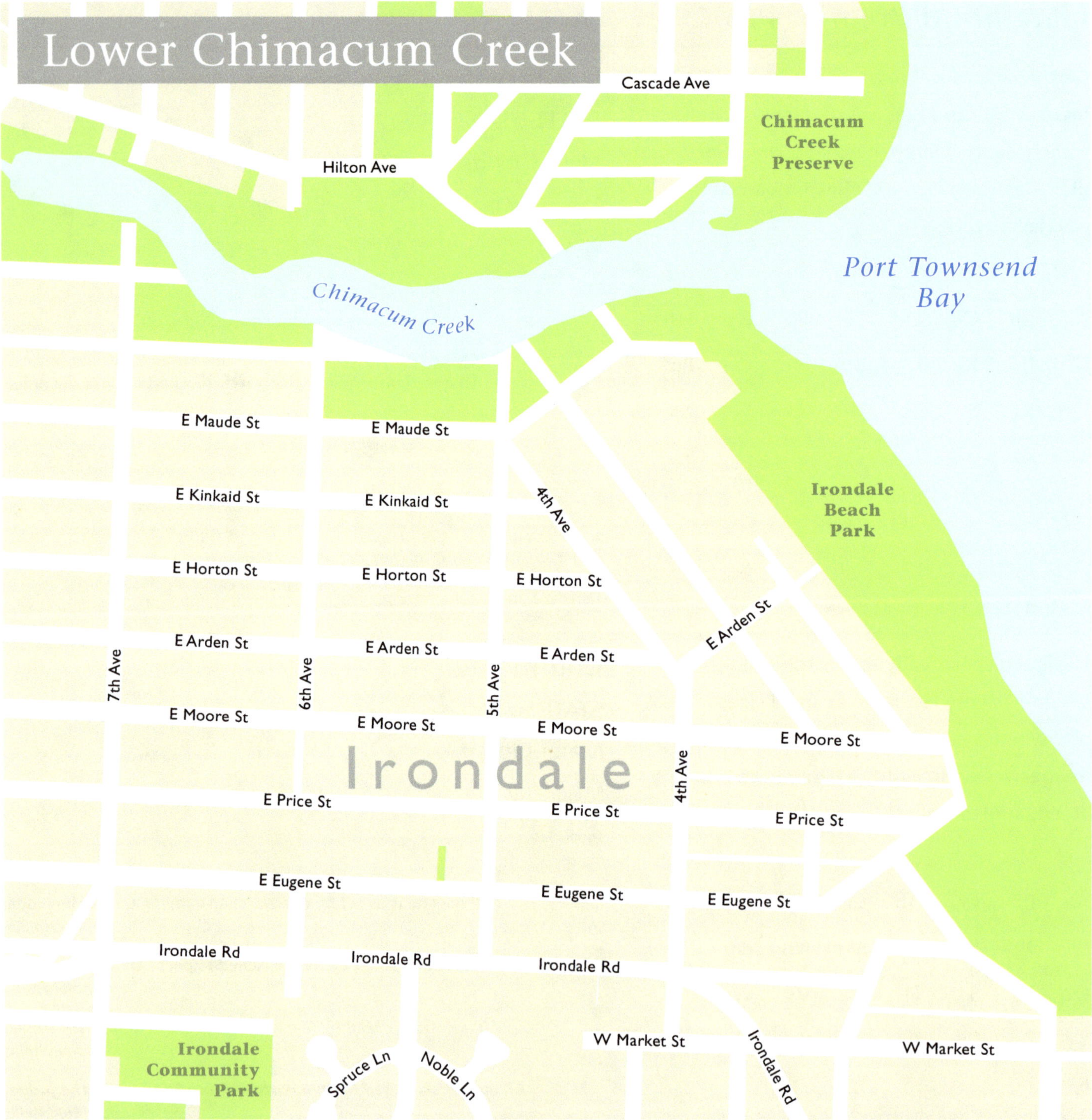

Lower Chimacum Creek

Cascade Ave
Chimacum Creek Preserve
Hilton Ave
Port Townsend Bay
Chimacum Creek
E Maude St
E Maude St
E Kinkaid St
E Kinkaid St
4th Ave
Irondale Beach Park
E Horton St
E Horton St
E Horton St
E Arden St
E Arden St
E Arden St
E Arden St
7th Ave
6th Ave
5th Ave
E Moore St
E Moore St
E Moore St
E Moore St
Irondale
4th Ave
E Price St
E Price St
E Price St
E Eugene St
E Eugene St
E Eugene St
Irondale Rd
Irondale Rd
Irondale Rd
W Market St
W Market St
Irondale Community Park
Spruce Ln
Noble Ln
Irondale Rd

ITC Stone Serif designed by Sumner Stone

ITC Stone Serif is part of a typographic family that also includes sans and informal fonts. This font is somewhat condensed and has a tall x-height. It is good for large and small text blocks and labels.

Country 16pt

STATE 12pt

Major City 12pt

Minor City 10pt

Town 8pt

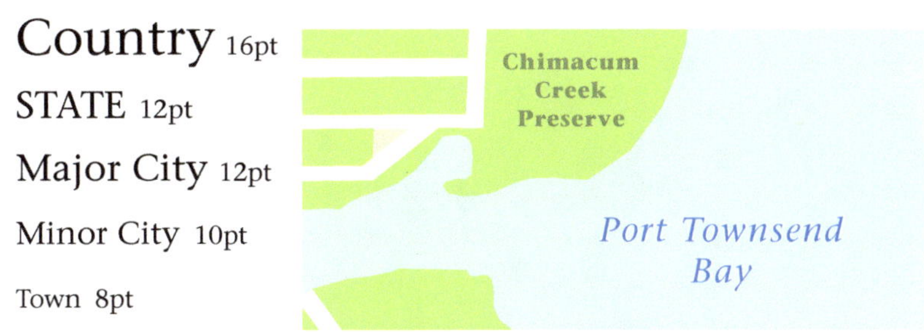

Aa	Bb	Cc	Dd	Ee	Ff	Gg	Hh
Ii	Jj	Kk	Ll	Mm	Nn	Oo	Pp
Qq	Rr	Ss	Tt	Uu	Vv	Ww	Xx
Yy	Zz	0	1	2	3	4	5
6	7	8	9				

All questions asked by five watched experts amaze the judge.
ITC Stone Serif Medium

All questions asked by five watched experts amaze the judge.
ITC Stone Serif Bold

All questions asked by five watched experts amaze the judge.
ITC Stone Serif Medium Italic

All questions asked by five watched experts amaze the judge.
ITC Stone Serif Bold Italic

Gill Sans designed by Eric Gill

Gill Sans is an arty looking sans serif that has some very wide and some very narrow capital glyphs that harmonize when combined, making it a nice display font. It is often included in Mac OS™.

Country 16pt

STATE 12pt

Major City 12pt

Minor City 10pt

Town 8pt

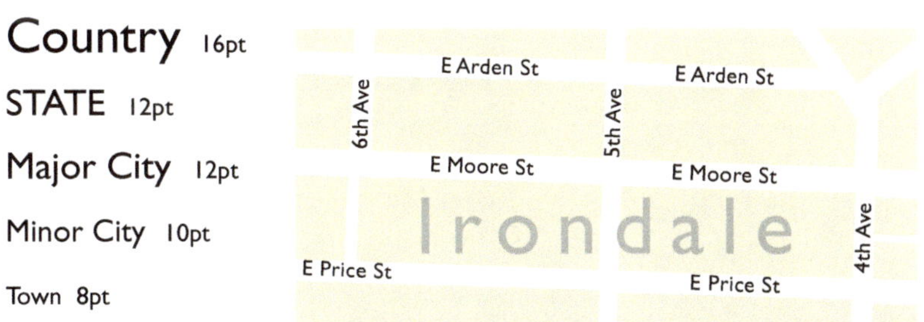

Aa	Bb	Cc	Dd	Ee	Ff	Gg	Hh
Ii	Jj	Kk	Ll	Mm	Nn	Oo	Pp
Qq	Rr	Ss	Tt	Uu	Vv	Ww	Xx
Yy	Zz	0	1	2	3	4	5
6	7	8	9				

All questions asked by five watched experts amaze the judge.
Gill Sans

All questions asked by five watched experts amaze the judge.
Gill Sans Bold

All questions asked by five watched experts amaze the judge.
Gill Sans Italic

All questions asked by five watched experts amaze the judge.
Gill Sans Bold Italic

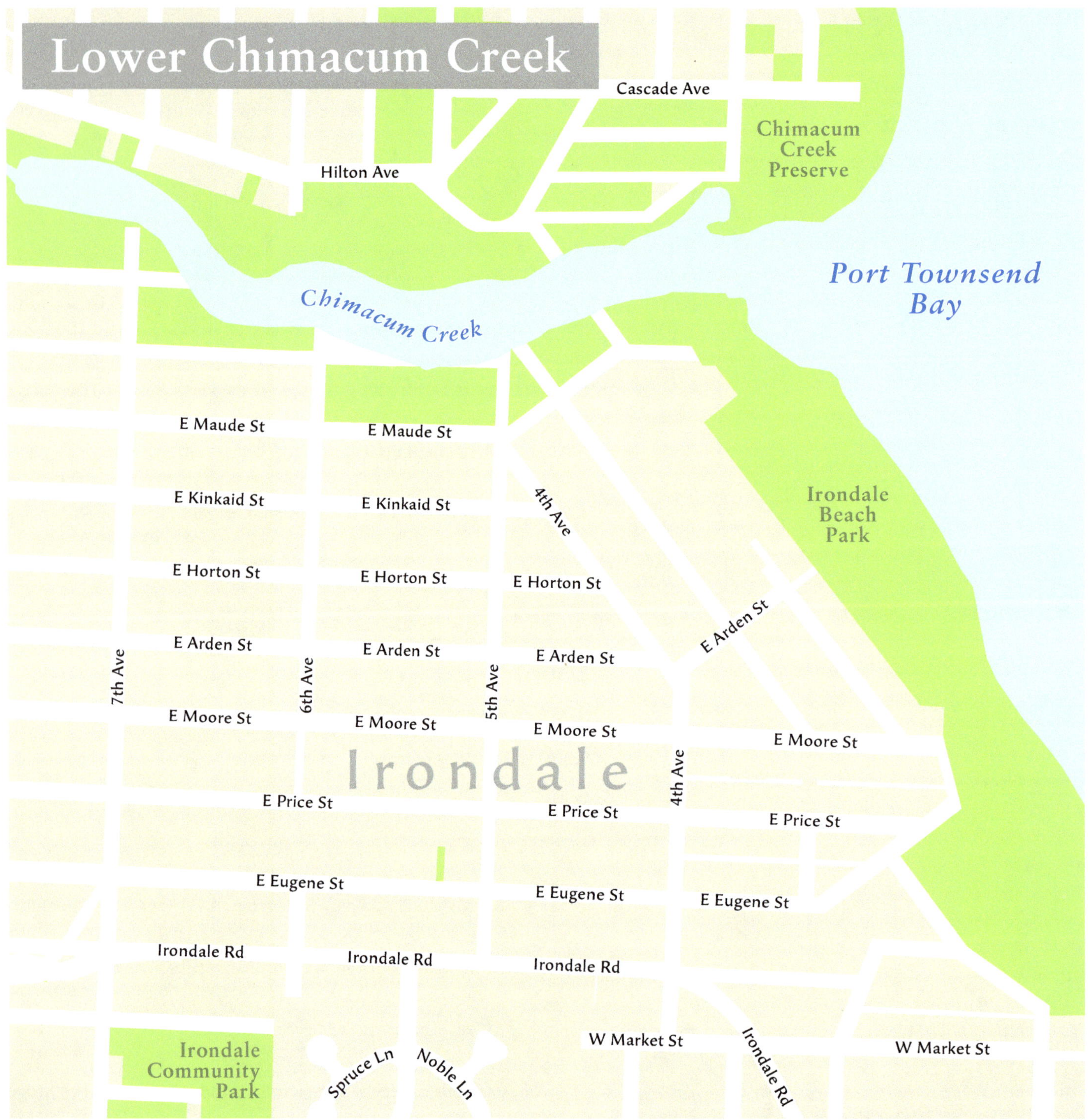

Lower Chimacum Creek
Cascade Ave
Chimacum Creek Preserve
Hilton Ave
Port Townsend Bay
Chimacum Creek
E Maude St
E Maude St
E Kinkaid St
E Kinkaid St
4th Ave
Irondale Beach Park
E Horton St
E Horton St
E Horton St
E Arden St
E Arden St
E Arden St
E Arden St
E Arden St
7th Ave
6th Ave
5th Ave
E Moore St
E Moore St
E Moore St
E Moore St
Irondale
4th Ave
E Price St
E Price St
E Price St
E Eugene St
E Eugene St
E Eugene St
Irondale Rd
Irondale Rd
Irondale Rd
W Market St
W Market St
Irondale Community Park
Spruce Ln
Noble Ln
Irondale Rd

Sabon designed by Jan Tschichold

Sabon, a Garamond derivative, is sophisticated yet different. The letter-forms are complex and reminiscent of classical calligraphy and have a bold and stately personality. It is highly readable due to its tall x-height.

Country 16pt
STATE 12pt
Major City 12pt
Minor City 10pt
Town 8pt

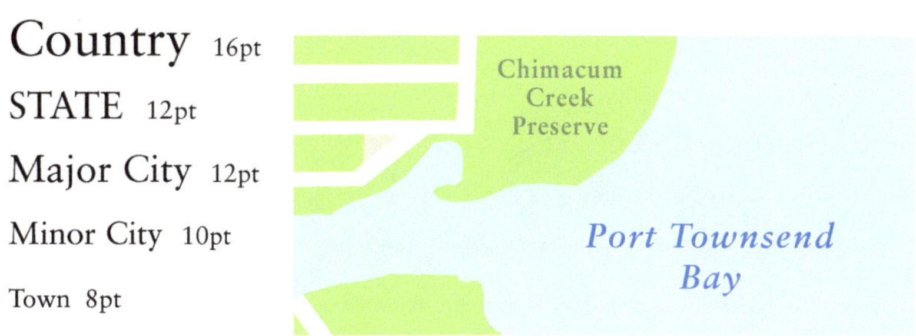

Aa Bb Cc Dd Ee Ff Gg Hh
Ii Jj Kk Ll Mm Nn Oo Pp
Qq Rr Ss Tt Uu Vv Ww Xx
Yy Zz 0 1 2 3 4 5
6 7 8 9

All questions asked by five watched experts amaze the judge.
Sabon

All questions asked by five watched experts amaze the judge.
Sabon Bold

All questions asked by five watched experts amaze the judge.
Sabon Italic

All questions asked by five watched experts amaze the judge.
Sabon Bold Italic

Beorcana designed by Carl Crossgrove

Although Beorcana is a sans serif, it is still quite readable at small sizes in large text blocks. Its unusual appearance is calligraphy-inspired, imbuing text with a lively personality.

Country 16pt
STATE 12pt
Major City 12pt
Minor City 10pt
Town 8pt

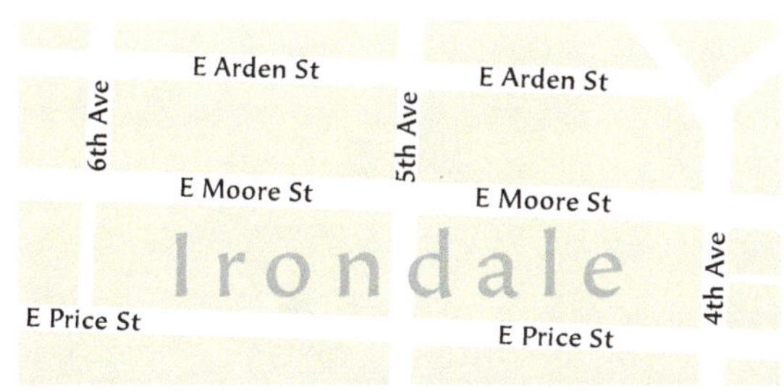

Aa Bb Cc Dd Ee Ff Gg Hh
Ii Jj Kk Ll Mm Nn Oo Pp
Qq Rr Ss Tt Uu Vv Ww Xx
Yy Zz 0 1 2 3 4 5
6 7 8 9

All questions asked by five watched experts amaze the judge.
Beorcana

All questions asked by five watched experts amaze the judge.
Beorcana Bold

All questions asked by five watched experts amaze the judge.
Beorcana Italic

All questions asked by five watched experts amaze the judge.
Beorcana Bold Italic

COMPOSITION PATTERNS

Composition Patterns

When beginning the design of a new map the cartographer can seek inspiration from existing maps, taking note of their overall characteristics as well as individual pieces or particular techniques, and build on that inspiration. For example, a cartographer tasked with mapping the journey of a 19th century explorer for a print publication should look at maps published in history books of a similar era to note everything from the overall look and feel to the inclusion or exclusion of north arrows, label fonts and weights, and line styles. From this information, the cartographer can begin to formulate many aspects of the map. This leads to a better base from which to ask the audience or client pointed design questions that will further improve the design.

This chapter provides some of the inspiration that the cartographer needs when beginning a new map design. It focuses on patterns of map-making that illustrate common cartographic solutions to common cartographic problems. These are observed techniques applied by cartographers in today's publications and applications. They are particularly relevant to the cartographer seeking to create high-level communicative design. Some of the patterns, like the flow arcs pattern, have just been recently invented, while others, like the small multiples pattern, have been in use much longer. All can be applied to both print mapping and interactive mapping applications.

In this chapter, the term **pattern** refers to a promulgated cartographic technique that is reusable, customizable, and proven to be effective. This is the same sense of the word *pattern* as used in the software design field, where design patterns allow for more efficient programming, since practitioners are not spending time reinventing solutions that others have already designed. The term **pattern** is used in this chapter exclusively in the sense described above, not to mean a regular or repetitive form in data. The patterns are loosely ordered from simple to complex. The first pattern, *signal-noise ratio*, explores the concept of minimizing all extraneous details to create a simple and clear map composition. The last pattern, aptly named *complexity*, discusses the idea that many of the most successful map compositions are also the most complex. The other patterns are arranged in between these two.

Beginning cartographers are advised to learn these patterns. Beginning cartographers should also learn the basic principles of cartography, which are not covered here. The advanced cartographer may or may not be initially familiar with the patterns described here. A cartographer might use two or even three of these patterns together to present a particular set of spatial data in the best way. Therefore, it is recommended that the reader make an effort to read the whole chapter thoroughly so that no potentially applicable pattern will be overlooked. A thorough reading will also help if more information is sought about concepts that appear in more than one pattern: creativity, complexity, and relief, to name a few. The chapter can also be skimmed and used as a reference, once this initial read-through is completed. By becoming expert at the patterns presented here—and in future publications on the subject—the cartographer will be able to decrease design time and effort while increasing quality.

An argument may be made that cartography is inherently a creative exercise, and as such, can't be cataloged and placed into discrete patterns as this chapter attempts to do. However, it is not the author's intention to suggest that these patterns replace creative thinking. Rather, the patterns should be used to enhance creative thinking by providing inspiration and springboards to better mapping. These patterns provide tested and proven techniques that prevent the time-consuming work of starting from scratch, but do not preclude the development of new techniques.

The patterns presented here are not comprehensive of every solution available to the cartographer. Indeed, a cartographer should definitely not feel restricted to only these patterns if a different and better method presents itself. Instead, they serve as a starting point for what the author believes will become a new field of research and publication in cartography: map composition patterns. Software design patterns are well documented in the software literature and are also accessible via interactive websites for easy reference. Likewise, these map composition patterns can be the foundation for others to add to, build upon, and enhance in order to promote exceptional cartography among practitioners.

Do Not Reinvent the Wheel *Why spend time inventing a wheel when it has already been done? Instead, you can spend your time designing a better wheel, drawing inspiration and knowledge from the initial design. You need to know that the wheel already exists and how it functions in order to make it better. In short, you want to have that "wheel" in your toolbox. Even if the "wheel" has to be reworked, it is helpful to have the initial design to start with. As software engineer Jim Humelsine states, "With design patterns I'm able to design more in less time with higher quality."*

Signal-Noise Ratio

USAGE The signal-noise ratio[1] is a formula that describes the amount of ink in a map that is used to represent data directly—signal—versus the amount of ink that represents nothing—noise. At first glance, it might seem ridiculous that a cartographer would put ink on a map to represent nothing, but in fact, it does happen. Obviously, the signal portion of the equation is maximized on effective maps while the noise portion is minimized.

Improvement of signal-noise ratios is seen in iterative works. For example, the elements that contribute to noise may be published in a first edition but removed in subsequent iterations once it is clear that they provide undue clutter. Perhaps the cartographer initially places frames around the legend, title block, byline, and scale bar, but after a time realizes these do not contribute anything to the map's message. Perhaps the cartographer initially places boundaries around all the nested sub-basins in a watershed map, but upon further reflection, determines that the sub-basin boundaries can be indicated with a less-obtrusive area-fill instead.

This is usually summed up as: if it doesn't add anything worthwhile to the map, get rid of it. However, this general rule is quite a bit more complicated when we consider how it can apply to maps that use a lot

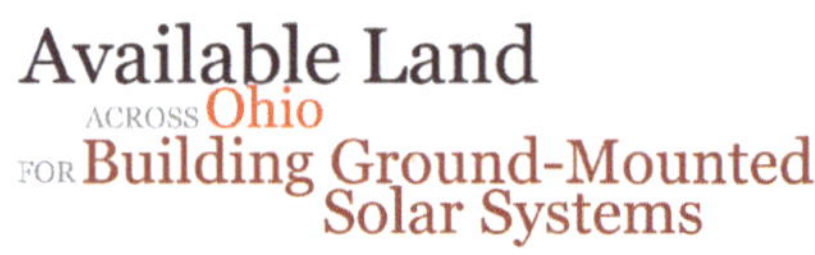

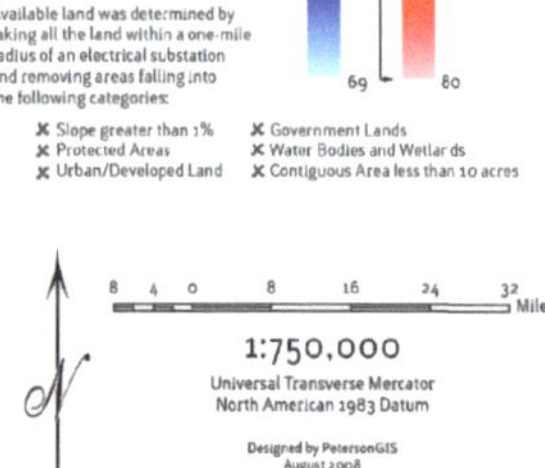

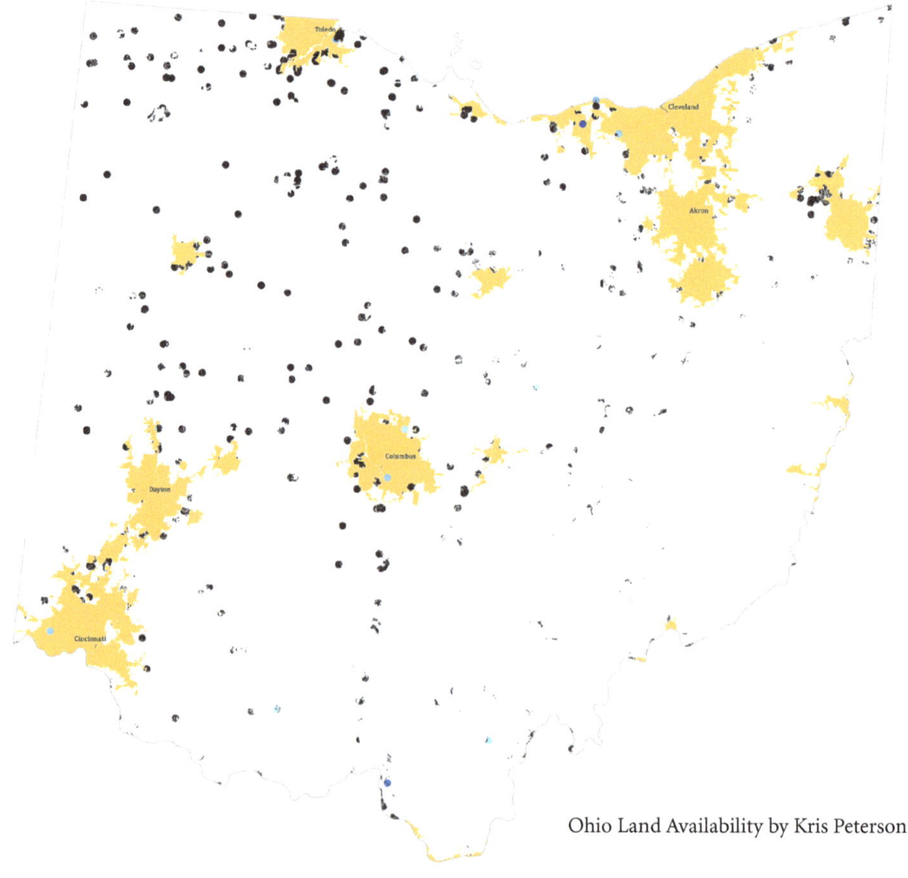

Ohio Land Availability by Kris Peterson

of flourish. In classic maps, for example, seemingly extraneous details abound—but the maps are still eminently readable. The signal-noise ratio rule, then, should be implemented as follows: take a critical look at the map features and elements to determine if they are interfering with the main point, and if so, toss them.

STYLE Maps with maximum signal-noise ratio may have a lot of white space in them, especially if the map product is for displaying conclusions or presenting analyses to a limited group of stakeholders. These maps do not require detailed backgrounds and can instead simply display the salient information. Another signifier of this style is that it has clear-but-unobtrusive boundaries between features, or, where they are not needed, no boundaries at all. These maps may have only basic, non-textured backgrounds or whitespace backgrounds. Flow arcs—where the only connections between points are displayed and where these connections form a pseudo background reflexively—have the highest signal-noise ratio possible.

TECHNIQUE For maximizing the signal part of the equation, the cartographer must ensure that all information displayed on the map is either of primary relevance to the map reader

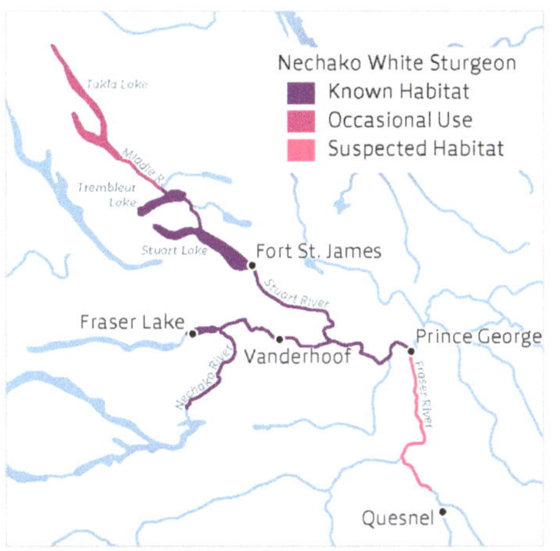

Nechako White Sturgeon by Eliana Macdonald, Ecotrust Canada.

or of a nature that provides clear, uncluttered, and necessary context (see also Complexity). To minimize the noise part of the equation, eliminate anything that interferes with the clarity of the composition. This seems simple enough, but in fact, the main problem one has when perfecting a map with a high signal-to-noise ratio is identifying which elements are only providing noise, since the noise is often not immediately recognizable as such to the cartographer who designed it. This is where peer review can be helpful.

The flourishes on maps of a classic style do not typically *interfere* with the reading of the map. Likewise, the items on a good piece of modern cartography should not interfere with the reading of the map either. For example, if polygon features can be grouped with nearby features that have the same values, the polygon boundaries may become unnecessary visual clutter. While the cartographer's default position is to show those polygon boundaries (because that is the way the underlying data is organized), it does not take much additional effort to group them into a more appealing configuration.

[1]A related concept is Tufte's term data-ink ratio. Tufte, Edward R. *The Visual Display of Quantitative Information* (2nd ed.). Cheshire, CT: Graphics Press, 2001.

Discontinuous Frame

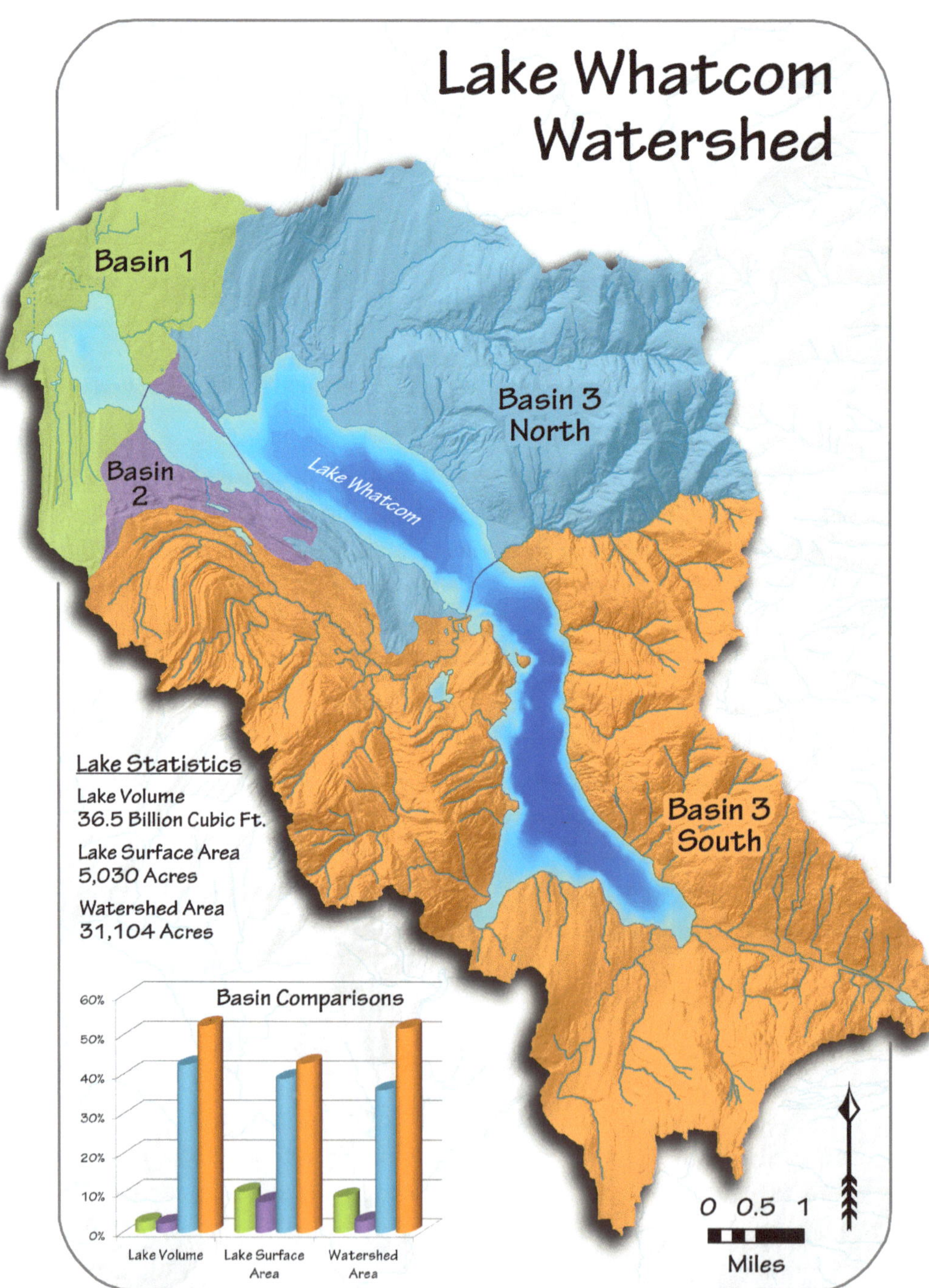

Lake Whatcom Basin Comparisons by Chris Behee, City of Bellevue.

USAGE A discontinuous frame is a simple graphic technique whereby the frame (border) of the map page is "broken" in one or more areas, allowing the map to jut out. This is useful when a map's region is a peninsula or circular shape such as, for example, a map of Italy or a globe. This pattern reminds the cartographer to think outside the box, literally, in order to fit one or more elements into a predetermined space or to add additional visual interest.

STYLE Two ways of interrupting the border graphics are fading and abrupt stops. In fading, the frame gradually diminishes where the map encroaches on it. In abrupt stops, the frame simply ends where the map touches it, or ends just before the map touches it. The amount of space left between the encroaching element and the frame is left to the cartographer to determine, usually via trial and error. Elements other than the map can also bleed into the frame, such as illustrations, titles, and legends.

Some 3D maps use this technique to further emphasize a portion of the map that is particularly high, such as a rendering of Mount St. Helens or a thematic 3D map of the largest cities in the world. In these cases, the visual suggests that a part of the map is so high it could not fit onto the map page.

TECHNIQUE It is easiest to accomplish this pattern by moving into a graphics program, though certain clumsy workarounds may exist in traditional mapping software through the use of graphic elements, such as rectangles, that match the background color to overlap and obscure parts of the frame. Only one or two frame discontinuities should be used per map page, otherwise the frame appears too irregular. It is recommended that the cartographer be aware of other graphic design techniques, because this is just one of many ways to create a well-designed page layout. Consult graphic design literature to learn more.

Flourish

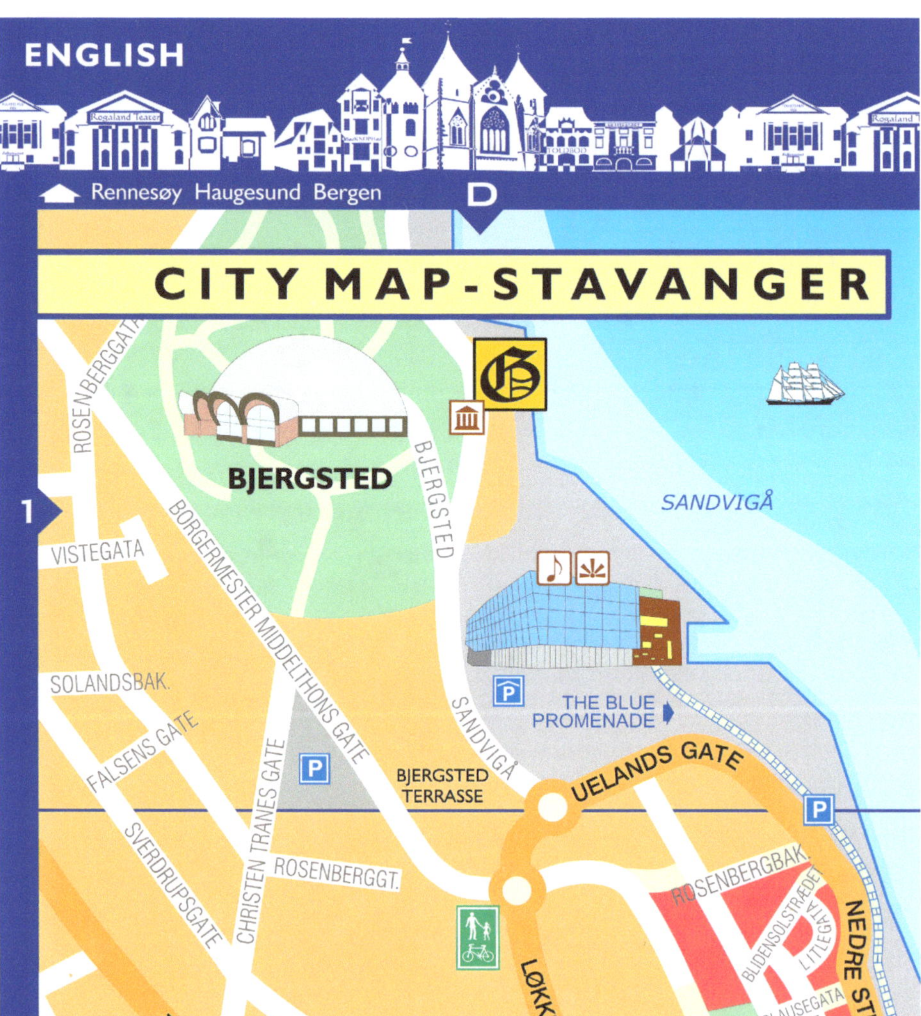

Stavanger by Kevin-Paul Scarrott, Stavanger Guide Maps Norway.

So Geographers in Afric-Maps
With Savage-Pictures fill their Gaps;
And o'er unhabitable Downs
Place Elephants for want of Towns.

Jonathan Swift, "On Poetry: A Rapsody"

USAGE Flourishes are ornaments placed on a map to fill extra space or to add decorative flair; they typically have no other usefulness aside from the aesthetic. Any map with content that is quaint, kitsch, or pictorial—such as children's art or historic-looking maps—can benefit from the use of flourish. It is uncommon to find flourishes on maps created by those in the geosciences, as map products from that field are created to explain rather than to engage (though these strategies are not necessarily at cross-purposes). Classic maps from previous centuries tended to include a lot of flourish, especially prior to the 19th century. In modern maps, flourishes are still found in information graphics maps, gaming maps, fiction illustrations, and many other types of maps.

STYLE Flourishes of myriad types can, and have been, employed in innumerable ways. Cartographers trying to mimic historic-looking maps may include flourishes such as faux rag edges, images of dragons or people in vintage costume, banners, or detailed feature renderings such as waves or mountain peaks. Indeed, in classical maps, it seems the cartographers had no end of creative ideas for their maps. For example, the scale bars in several 17th century maps are accompanied by sketches of drafting compasses, with one tip on one end of the scale and the other tip on the other end of the scale.[1]

Typefaces that have a lot of flourish, such as old-time Garamond,

Caslon, Baskerville, or Fell, go well on maps that also contain other flourishes. For even more flourish strength, pure calligraphy typefaces such as Cygnet, Mutlu Ornamental, or Adine Kirnberg Script are used, especially on map titles.

Cartouches are a specialized type of flourish that were used extensively in classic cartography and can sometimes be found on modern maps that seek to mimic that look. They are typically quite large in comparison to modern flourishes, in that they may take up one-sixth or so of the map page. A pedestal topped with the cartographer›s portrait, an emblem or shield, and other sketched drawings are all common in cartouches. Calligraphic writing, generally large in size, is used on these cartouches to display the map title, cartographer's name, and date. Maps prior to the 18th century have more intricate cartouches than the ones found in the late 18th century and beyond.

More modern flourishes include illustrations of the map subject—sketches of dolphins surrounding a map of dolphin movement, for example. Modern flourishes can also include decorative borders or specific subject-appropriate details on children's maps, such as a pirate skull on a treasure map.

TECHNIQUE While modern map design tends toward the clean and simple, with white space being courted rather than shunned, flourish-style maps are the opposite. In these, the cartographer will want to fill most areas with map elements, graphics, or supporting text. The trick with these is to ensure that the flourishes are numerous and well distributed across the page. If only one or two flourish elements are on a map, they will appear superfluous.

In mapping software, illustration flourishes are typically added by using a marker symbol on a randomized point feature created for the task. Remember that marker symbols can also be used to represent things such as mountain peaks, waves, groves of trees, and other items. Alternatively, single flourishes, such as a drawing of a cruise ship, can be added as graphics to the final layout. Typographic flourishes are either hand-sketched—a drawing tablet computer is best for this—or typed in a text box after installing the typeface on the device. If the latter technique is used, be sure to embed the font in the finished electronic map file.

See also: Sketch, Illustrated

[1]See map titled "A Description of East India" by William Baffin and Thomas Roe, 1619, for one example.

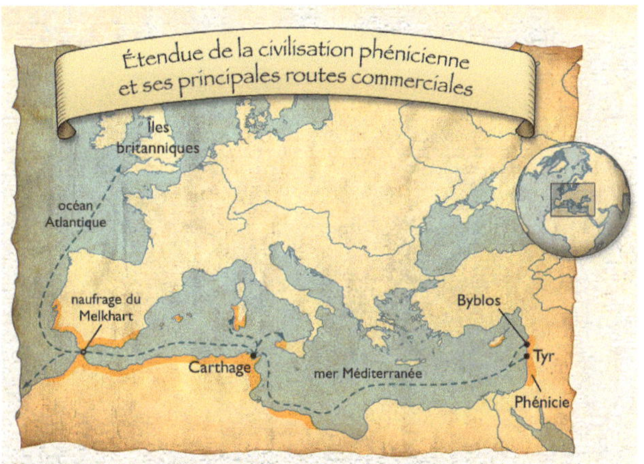

The Phoenician Civilization and its Main Commercial Routes by François Goulet, fg cartographix. Reproduced with the permission of QA International, from the book "Trésors enfouis". © Les Éditions Québec Amérique inc, 2007. All rights reserved.

Organic Arrangement

USAGE The organic arrangement pattern is used when there are several-to-many ancillary items that need to be placed in, around, or on the main map on a page. These might include photographs, illustrations, text boxes, inset maps, or all of these. While the cartographer's natural tendency could be to place the items in a separate portion of the layout, such as a sidebar—and this is entirely acceptable in many cases—the point of discussing this pattern is to remind the cartographer that it is also quite effective to integrate

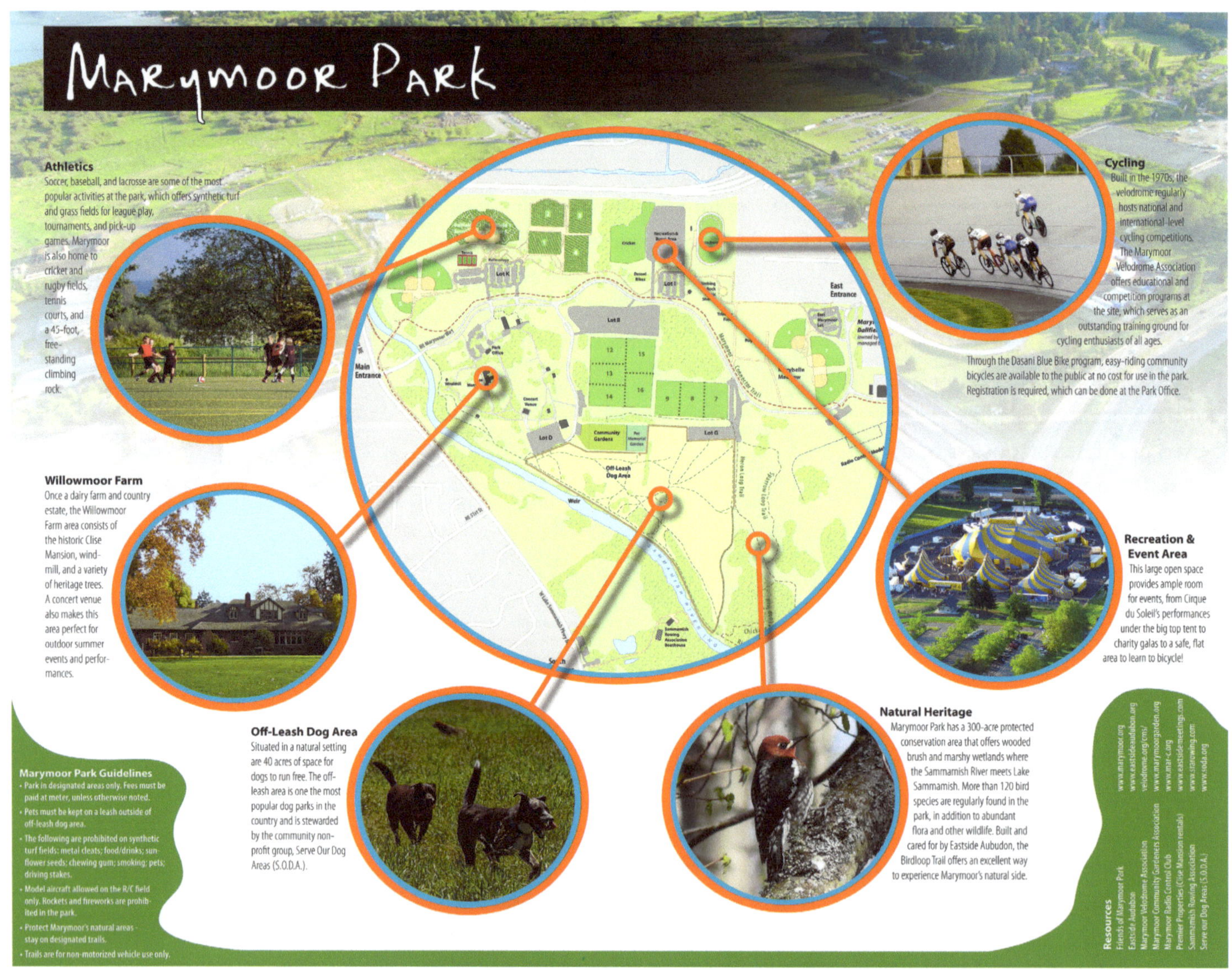

Marymoor Park by Matt Stevenson, CORE GIS LLC.

the items, almost as if they were map layers themselves.

This pattern commonly emerges on static layouts, especially folded printouts meant as navigational aids, store locators, and the like. Additionally, it is used in maps that have uniquely shaped geographies. It allows the map to be as large as possible, at the full extent of the map page. Maps meant for navigating a city's streets, where the city wraps around a large body of water such as a lake or canal, are suitable for this. The map can be maximized while still allowing, for example, an index and authorship information to be shown by placing those items on the water areas.

STYLE In organic arrangement, instead of lining up the map elements in discrete rows and columns, the ancillary map items are placed where they make the most sense, given the underlying main map. If the map, for example, is of a peninsula, the ancillary items may be arranged around it in the water area. The items don't always have to be placed on a water area; they can be placed on whatever areas of the map are the least important to the viewer. The underlying map drives the placement of the items but does not preclude the use of an organizing construct—such as the circular nature of the photos in the map example in this section—as long as it does not interfere with the main map.

TECHNIQUE Identify the unimportant locations of the main map. Place the ancillary items on these areas, perhaps with drop-shadows, frames, or edge-glows to provide a visual disconnect from the map, or with fade-outs at the edges to allow them to blend in with the map while still providing some visual separation. Alternatively, fade out or clip off the background map where the ancillary items will go. Fade-outs can produce a more cohesive look than clipping. Sometimes no separating mechanism is needed, especially when the item has high contrast with the underlying map data, such as an index with a white background layered on a dark blue water area.

Maintain visual balance by trying to place heavier-looking items opposite dense parts of the map. When the ancillary items are small or light, place them in groups to increase their visual weight. Groups with odd numbers of items generally look better than groups with even numbers of items. Give a thought to visual hierarchy as well, by placing ancillary items more toward the bottom and right side of a map rather than toward the top or left side of the map, if possible. If the items are of equal or higher importance than the map, then obviously they can be placed in whatever location works best.

If necessary, provide connecting lines between the ancillary items and their locations on the main map via simple, straight lines in a neutral hue. Make many iterations of the arrangement before deciding on final placement. Remember, the ancillary items do not have to be rectangular; they can also be circular or amorphous depending on their own underlying shape.

Vignettes

USAGE Vignettes provide extra contrast along coastal map features. They are often found in historic maps and historic-looking maps; though they are also employed frequently in medium-scale contemporary maps where extra contrast may be important enough to warrant the effect. In contemporary maps, they provide an extra element of artistic richness that adds to the overall design aesthetic by avoiding what can otherwise look bland on medium-scale maps.

STYLE Vignettes come in a few varieties including wavy or curved lines, repetitive triangular shapes, and color-fading or edge-glow effects. The commonality is that the lines, shapes, or color effects start out intense near the coastline and gradually decrease in intensity as distance increases perpendicularly from the coastline. This intensity may come from the color, the line weight, the line style, or the distances between the vignette features, or all of these.

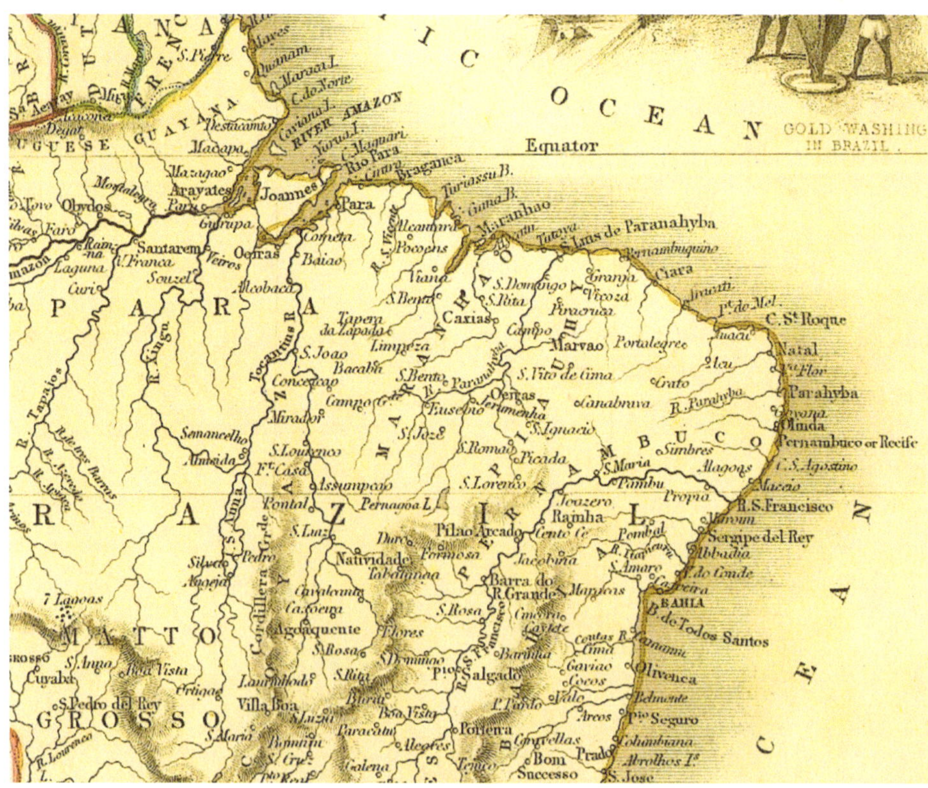

Vintage Map of South America by J. Tallis and Sons, 1851.

Commonly, the color remains constant, or at least similar, but the distances between the vignettes increases from almost touching near the coast to slightly further apart before ending. The lines can also become fragmented as the distance increases, creating a look similar to waves gathering momentum as they near the shore. The color of the vignettes is often just a few shades darker than the water fill. Shading or edge-glow vignettes are generally found on contemporary maps, whereas concentric line vignettes are more indicative of historic maps.

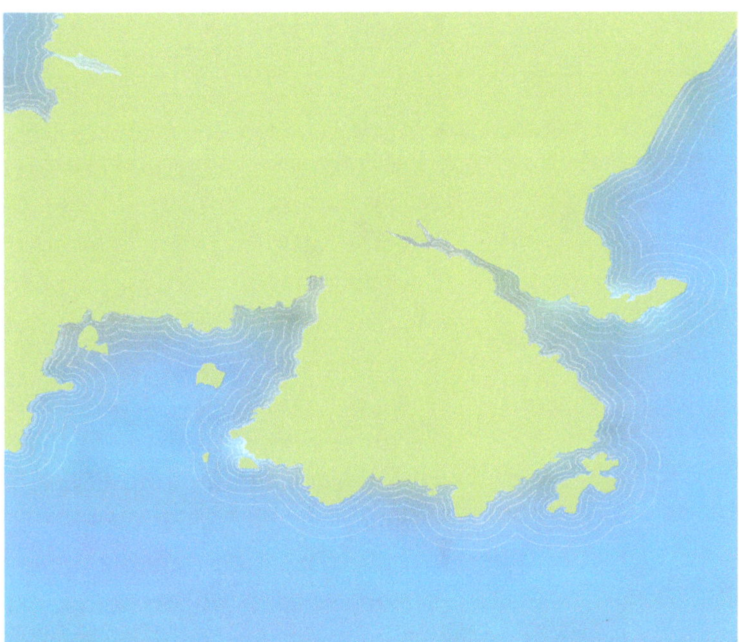

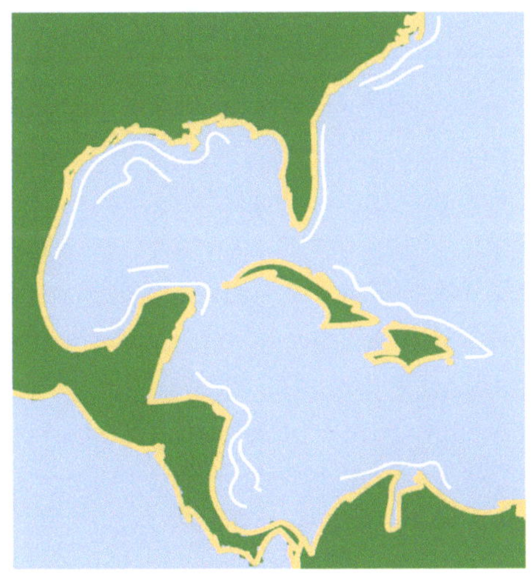

by Gretchn N. Peterson

TECHNIQUE In mapping software, a Euclidean distance function can create the needed shading or glow effects; the output is shaded according to distance from the shoreline, which is represented in a raster dataset. The raster data need to be created at a fine enough resolution to avoid a pixilated look. Some trial and error will be required when choosing the color ramp for the finished raster. In graphic design software, a glow effect can be created to provide a similar look.

To create concentric line vignettes, outside contour lines can be created with a buffer operation, using the coastline as the starting point. Whichever vignette technique is used, the amount of distance that the lines span will depend on the map scale and the preferences of the cartographer. They can range from quite narrow to filling up the entire body of water, prior to starting again at another coast on the opposite side—for a large bay, for example.

Focal Point

USAGE A focal point is a deliberate visual emphasis on one aspect of a map, usually to focus the map reader's attention on a particular region in the map. This pattern is often used as part of an overview map, to highlight the part of the larger region on which the map is centered. It can also be used anywhere an extra amount of emphasis is needed but where it isn't feasible or desirable to completely clip out the surrounding features, if, say, these features can provide some context without competing with the highlighted feature.

STYLE The most common ways that cartographers create focal points are by color and by masking. To highlight by color, the cartographer creates a high amount of contrast between the highlighted feature and the surrounding features—dark blue with cream, for example. Masking involves the lightening of surrounding features to focus attention on the non-lightened region by applying a semi-transparent feature layer on top of the non-essential parts, thereby lightening them. Less common ways to produce focal points are to offset them with a drop shadow (a localized 3D effect), to have them flash on and off in animated maps, or to create an edge glow around the features.

TECHNIQUE It is best to highlight for maximum effect. The novice cartographer's tendency is to under-emphasize a focal point. If a feature is going to be highlighted as a focal point, the emphasis needs to be clear and bold. For example, in a map of a proposed community park, the cartographer may err in highlighting the park area by using only a slightly darker outline or a tiny drop shadow. But a park proposal map should clearly define the boundaries of the proposed region by offsetting it in a bold way—for example, by both darkening its principal components and masking (i.e., lightening) the surrounding areas. For another example, in a trail map, the trails are of primary importance and should be much darker and bolder than other line features, such as roads and topography lines.

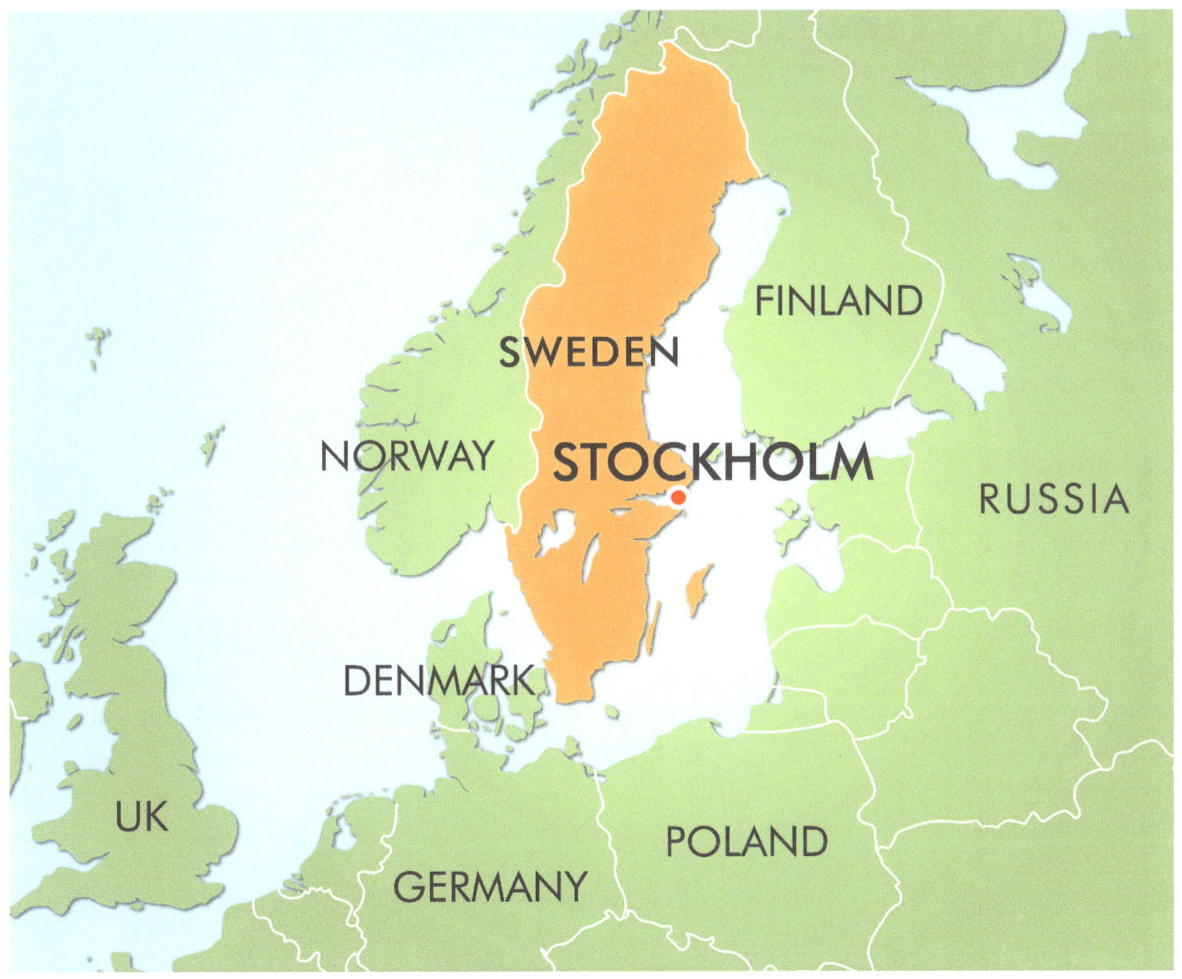

Stockholm, Northern Europe Map by Hugo Ahlenius, Nordpil.

Line Highlights

USAGE Linear features are often difficult to adequately highlight, due to their relative small size in comparison to the page. Obviously, at large scales, linear features can be turned into polygonal features—river lines to bank lines, road lines to road casings, and so on—which helps to emphasize them. But at smaller scales where this cannot be done, the feature must be highlighted if it is the central focus of the map. Line highlights are used, therefore, in all cases where a linear feature must take visual precedence in a map composition.

STYLE Some maps that highlight lines go the easy route by eliminating everything except the lines, line labels, and a few point features. On the opposite end of the spectrum are maps with very detailed backgrounds—ski maps, for example—where the lines are highlighted by means of both bright colors and vivid dash, dot, and dot-dash patterns. In some maps, the highlighted line is made up of two

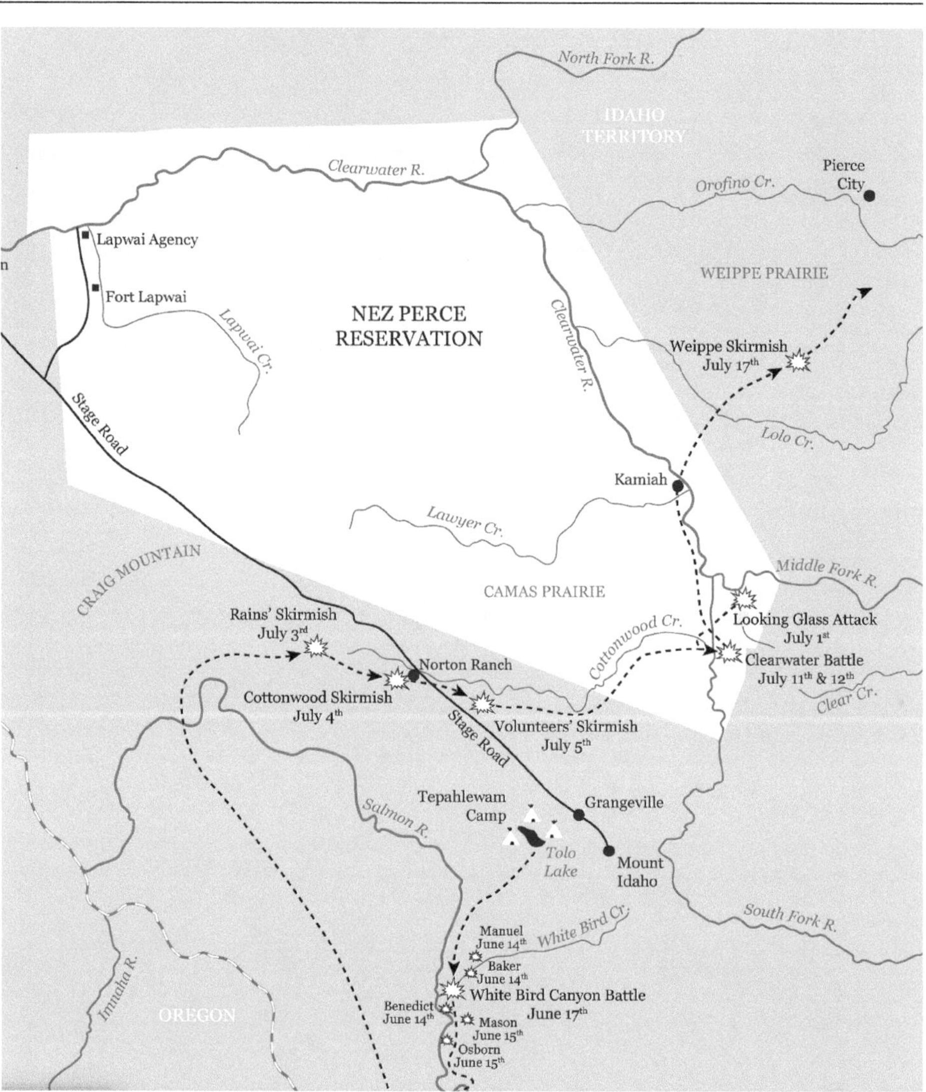

Nez Perce War by Kristian R. Underwood, Underwood Graphics.
West, Elliott. *The Last Indian War: The Nez Perce Story.* New York: Oxford University Press, 2009.

lines where a thinner (perhaps dotted) line is displayed in a bright color, which is further set-off by a darker, thicker line underneath. This dual-line style is especially effective when the underlying background varies in color because the line underneath provides an even, contrasting color for the top line.

TECHNIQUE Highlighting line features is difficult but not impossible. The first thing to try is mapping out just the lines and nothing else. If this were the entirety of the map, would it be adequate? Depending on the subject, it may be. If not, try widening the line to the maximum extent possible, while maintaining spatial integrity, especially around curves or intersections. Also advised is a double-color and double-symbol line feature (see Style, above). This is accomplished by creating a wider line underneath a thinner line, where the thinner line is a brighter, dashed line. If using a dashed line symbol, make certain that it will not be confused with any political borders that may be symbolized similarly. The political borders can be re-symbolized with solid lines, perhaps, in order to both lessen their visual impact and provide adequate feature distinction.

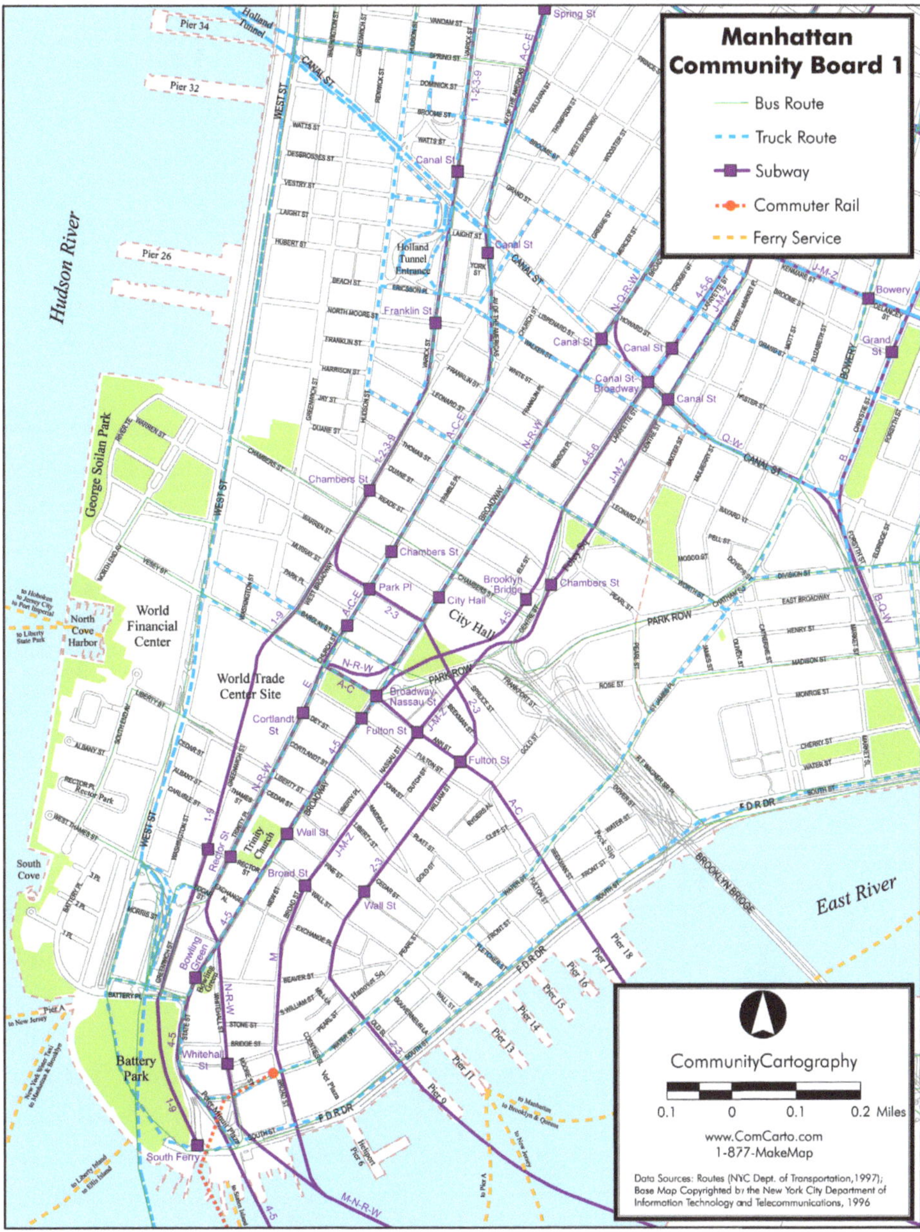

Manhattan Community Board 1—Transportation Networks by Danielle Hartman, CommunityCartography. © 2003 CommunityCartography, Inc.

Aerial as Context

USAGE Aerial photos are increasingly easy for general map audiences to interpret due to their ubiquity in webmaps. This makes them useful as background layers for webmaps and print maps. Much like topographic relief, aerial photos provide near-instant recognition of place. On webmaps, the map reader usually has the choice of backgrounds to toggle on or off: an aerial photo, a transportation layer, or both simultaneously, along with the other information that is potentially displayed on top. A downside is that the layer's transparency cannot be changed by the user.

In print maps, the use of aerial photos is not as common as the use of hillshading or other topographic techniques. In print maps, however, the benefit to using an aerial photo background is that other background layers, such as roads and other infrastructure, can be kept to a minimum since the aerial photo will supply that information (sans labels, of course).

Aerial photos that provide background context are seen in maps of all scales, though they are most effective at very large scales, where individual trees and residential roads can be discerned. When used at smaller scales, they may not provide the needed location specificity, and they may also compromise the clarity of the composition. For example, at the scale of a large town, the rooftops and roads are generally too small to be helpful. However, larger features such as parks and highways may still be visible, which to some is enough of a reason to use them.

Please note that this pattern focuses only on the use of aerial photos as background *context* layers. Aerial photos are also used extensively in maps where they are the main focus—such as volunteered geographic information, time change analysis, and before-and-after disaster visualization. Those uses are not discussed here.

STYLE An effective aerial photo background layer is rendered in unsaturated color or grayscale. Saturated colors typically look amateurish and compete with other layers. Aerial photos that are semi-transparent and layered on hillshades are sometimes used to give map readers an even richer sense of place. Another interesting aerial style is the use of a grayscale aerial photo that provides context, where the "top layer" map data is just a few splashes of color placed on the areas (buildings, usually) of interest. For example, this colorizing technique can effectively illustrate which buildings will be affected by an ordinance change.

TECHNIQUE It is possible to add labels to a print map with an aerial photo background. Unless the aerial is significantly lightened in a masking process, the labels will need to be white or some other hue of high contrast with the dark background. Some map services provide aerial-photo-compatible label layers, which are useful for this. If the number of labels is manageable, it might be best to create and place them manually.

An aerial photo background is more visually cluttered than a hillshade background. Therefore, it is just as imperative (as with hillshade backgrounds) to ensure that it does not compete with the top-layer map features. To do this, the top-layer map features can be rendered in bright colors that have high contrast with the aerial photo, or the aerial photo can be lightened, or both. Be forewarned: a major mistake that cartographers make is failing to lighten the aerial photo sufficiently. One exception is if the map is of a sufficiently large scale—an oblique aerial photo of a small lake, for example—it can be kept in its original dark colors, especially if the other map layers are minimal, with

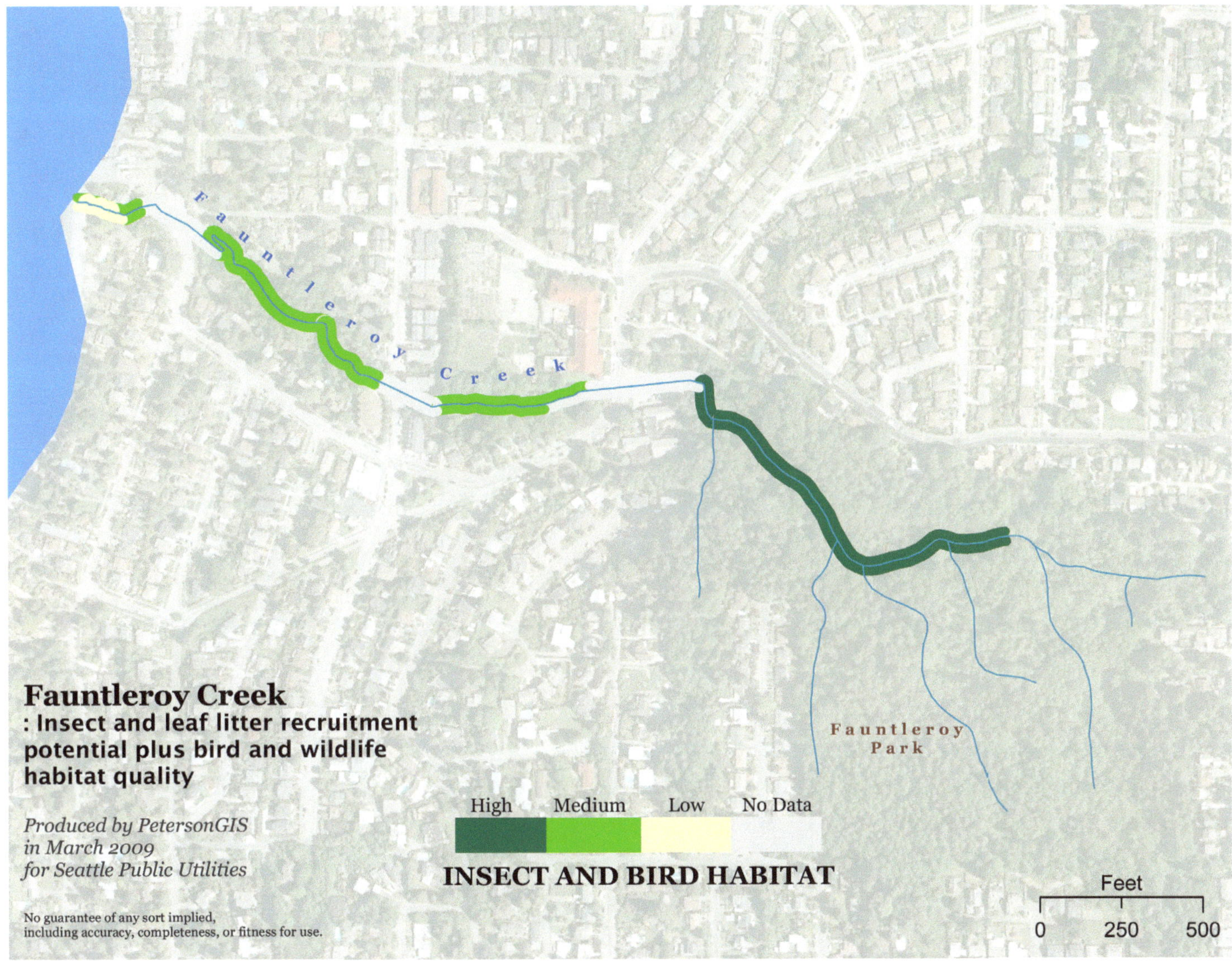

Insect and Bird Habitat Potential by Gretchen N. Peterson, PetersonGIS.

a few labels and arrows indicating lake homes, perhaps.

Colors used for the top-layer map features will appear differently when layered in a semi-transparent manner onto the aerial photo. Adjustments often need to be made to color choices once they are put in place. To further reduce contrast within the background aerial, take a look at the potential for simplifying the aerial with graphics software filters.

See also: Relief as Context

Relief as Context

USAGE It is often the case that providing shaded relief (terrain) context via a hillshade map background—with or without hypsometric tinting—gives the map reader who is unfamiliar with a region an immediate understanding of the location details. An argument against the use of relief as context is that it is superfluous clutter. However, if the hillshading does not overpower the map, it is usually advisable to include it. The pattern described here is about using hillshade as a contextual background element; not discussed here are cases where the terrain is one of the most important aspects of the map.

STYLE When hillshade relief is used solely for background information, it is usually subdued via masking or lightening. When large splashes of color are layered on top of it, such as, for example, in a choropleth map, the cartographer makes the top layer semi-transparent so that the relief is visible underneath. In these cases, the legend is matched to the map using color swatches (though some mapping software will now create legends that do this for you) since the original choropleth colors appear different when layered.

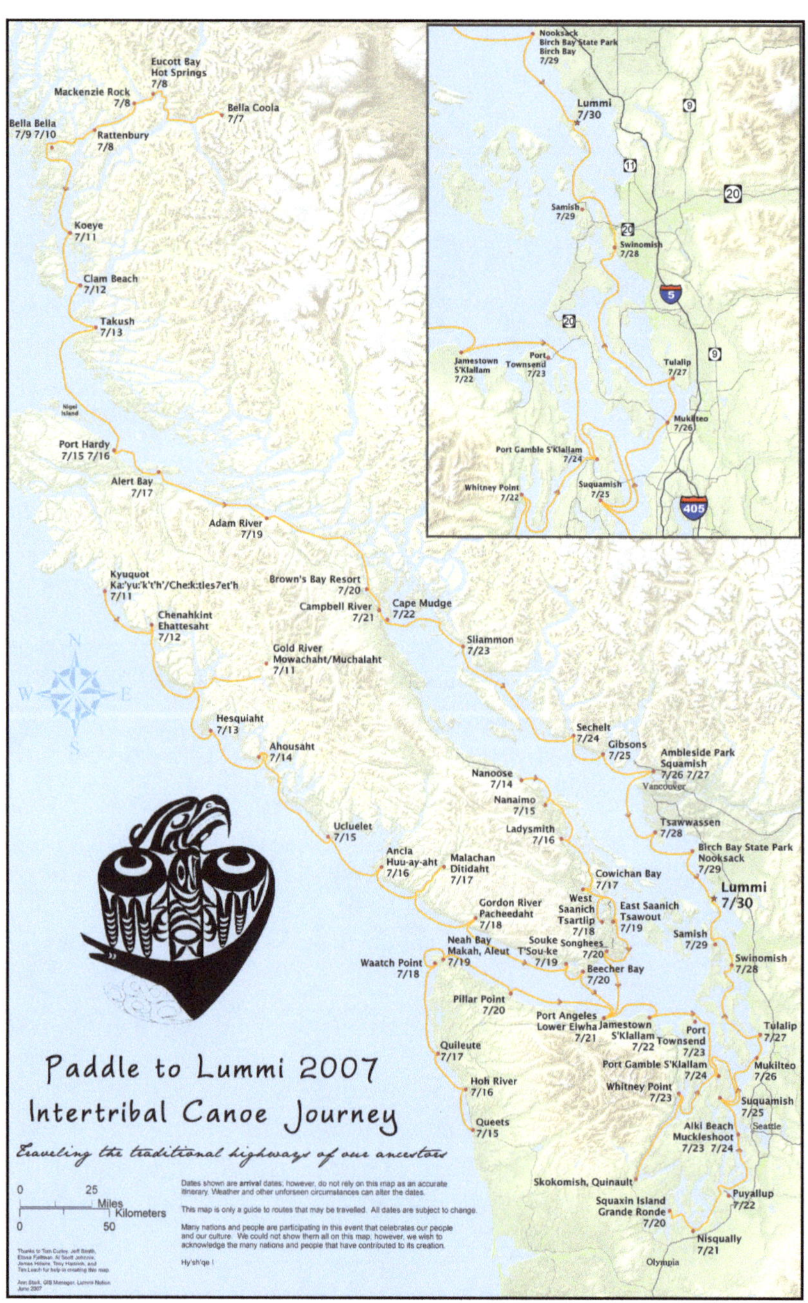

Canoe Journey by Ann Stark in collaboration with Tom Curley, Elissa Fjellman, and Tim Leach.

Sometimes different elevation datasets are used in different parts of the map, one more detailed than the other. For example, high peaks and low valleys or rocks and other features might be separated and given different amounts of detail via different resolution data. This creates a more pleasing contrast than might otherwise be found if the detail were constant across the composition.

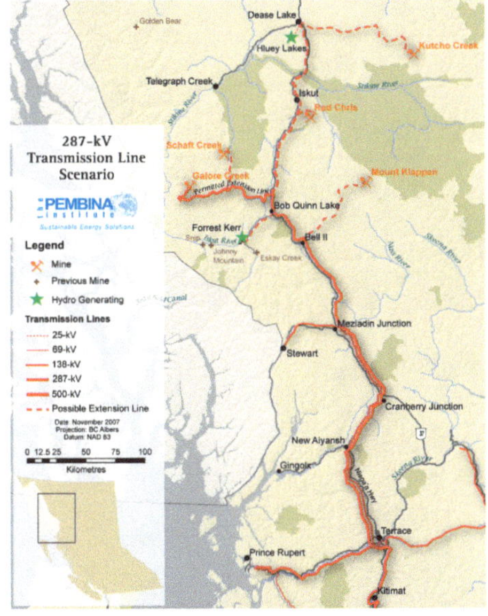

Transmission Line Scenario by Eliana Macdonald, Ecotrust Canada. Published in "Sizing it up: Scenarios for powering Northwest British Columbia." Karen Campbell and Greg Brown, The Pembina Institute. Feb. 2008.

TECHNIQUE It is important to pay close attention to the scale of the relief data. Many elevation datasets are too highly detailed for the scale that the cartographer needs. Down-sampling for small maps (e.g., a book map) coupled with exaggeration of the scale factor for the highest (or lowest) locations such as peaks and valleys may provide the context needed while decluttering the overall composition. Similarly, on interactive maps, the relief detailing should change depending on the user's zoom level. Because map readers can be led astray by pure hypsometric tinting—confusing the green used in lowlands as forest, for example—a cross-blended hypsometric tinting schema can be used instead. Cross-blended hypsometric tints are a mix of elevation and natural vegetation colors.[1]

If both high contrast and detail are needed in the shaded relief background, use graphic software tools to manipulate the tonal values. One aim of this procedure is to ensure that peaks, which should stand out to the map reader, have sufficient detail—something that is often not the case with computerized hillshading algorithms.[2] Another possibility is to combine two elevation datasets, one with a higher level of detail at the peaks and the other with a low level of detail in the valleys, into one elevation dataset prior to applying a hillshading algorithm.

[1]See Tom Patterson's Shaded Relief website for more information on cross-blended tints.

[2]This is Eduard Imhof's aerial perspective effect. Imhof, Eduard. *Cartographic Relief Presentation* (English language ed.). Berlin: Walter de Gruyter, 1965. (Reprinted by ESRI Press, Redlands, CA 2007).

Sketch

USAGE The sketch pattern includes whole maps or parts of maps that are hand drawn, or at least created to look as though they were hand drawn. They are often seen in literature to provide a visual of a fictional city or town. Sketch maps can also be used in any instance where a departure from the formality of computer-generated maps is warranted. Maps with sketch elements have turned up in the aforementioned fictional literature, on websites to accompany news items or by themselves, in artwork, and even on stage, sketched in front of an audience.[1] Note that, in this instance, "sketch map" is *not* used to mean a quick mock-up of a yet-to be-completed map.

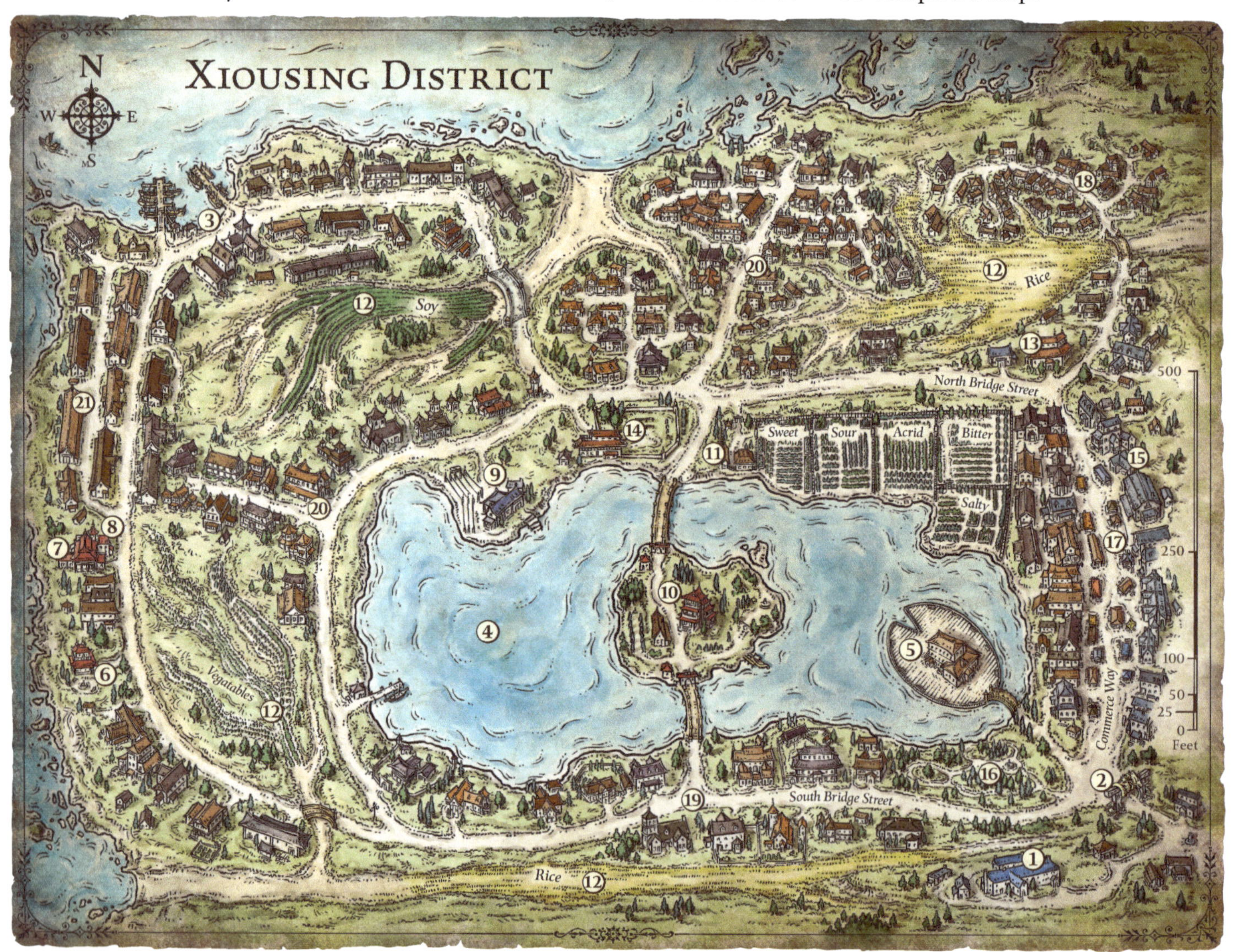

Xiousing District by Mike Schley, © 2011 Wizards of the Coast, Inc.

STYLE Sketch maps are often ornamental in style, but not necessarily so.[2] Some are highly illustrative, while others are simple direction-finding sketches. Sketched lines appear as though they were created with pen and ink; they often have irregular thickness and are not particularly straight. Many sketch maps are in black and white, but some are done in watercolor or other color medium. A map using this pattern may include only some sketch elements—a map of local sights drawn on a road basemap, for example—or may be entirely composed of sketching. Sketch maps tend to have personalities and are more artistic than computerized maps.

TECHNIQUE Because high-resolution graphics are needed for printing, maps for print publication are generally created via hardware such as a pen tablet and transferred to graphics software for finishing. Maps for digital media also benefit from the higher-resolution capabilities of tablet computers but are also occasionally drawn on paper and scanned, though this results in lower quality images.

The lines on sketch maps may be drawn in varying widths intentionally, to create the hand-drawn character that cartographers are aiming for with these types of maps. Interestingly, a steady hand, and therefore a steady line width, is one of the architect's hand-drafting skills. Sketch maps aim for the reverse. Here is a case where a map that may be drawn on a tablet computer is trying to appear a bit more relaxed, while it used to be that architects and other designers tried to make hand-drawn designs appear more rigid.

Particular attention to labeling is needed in maps with sketch elements. A map with only a few sketch elements may employ computer labeling (typography) effectively. However, a map that is entirely sketched will appear incongruous if the labels are not also hand-drawn. Even typefaces that mimic handwriting have no variation in strokes between instances of the same character glyph, making them less than ideal for the task. Zooming in to the drawing to create labels and other details is expected, but at a certain zoom level—300% or so—there is a risk of creating so much detail that the marks visually combine when viewed at the normal scale. Do some tests to determine the optimal sketching zoom level.

As with any map product, don't forget to consider the viewing angle for the map. Many sketch maps are bird's-eye views: an oblique perspective coupled with 3D techniques; while this requires more advanced drawing techniques, it is quite effective. A few mapping techniques are actually easier to accomplish in sketch maps than with completely computerized maps. For example, in classic cartography, shading is often created in water features and mountainous regions using a series of parallel lines that vary in distance from one another as a fill mechanism. They often run perpendicular to the feature that they are shading. While there are some ways to accomplish this technique in mapping and graphics software, it is a simple matter of sketching with a straight edge on hand-drawn maps.

See also: Flourish, Illustrated

[1]See the YouTube video titled "Senator Al Franken draws map of USA."

[2]See the Hand Drawn Maps Association online archive for inspiration.

Illustrated

USAGE Illustrated maps are primarily artistic devices that evoke a feeling about a particular place by use of graphic design elements that can be as unique as the artist creating them. They may be highly spatially accurate, or less so; they can represent real or fictional places. They may employ many of the elements of the sketch and flourish patterns, but not always. Illustrated maps do not provide exacting location awareness, nor do they maintain a consistent scale for all elements.

STYLE These maps are strongly graphically oriented. While some traditional cartography elements, such as roads and waterways, will usually be included, the main focal points are often fanciful drawings of individual places such as parks, historic homes, churches, and so on. The artwork is placed in its approximate location, though it is usually drawn much larger than the scale would otherwise justify. In this way, the map reader gets a feel for the important landmarks in a region. Often, such maps include nods to the way people interact with the space, through drawings of people, cars, boats, etc.

TECHNIQUE Bold color schemes are the norm for these maps, though that does not preclude the use of other palettes. One of the most important aspects of designing an illustrated map is deciding what the focal points will be (e.g., which parks, historic homes, and churches). If a real place is being mapped, the cartographer should either use her own sense of that place or, if the place is not personally known, should do extensive research to determine the appropriate inclusions and exclusions. Obviously, because the focal points are drawn at larger-than-life scale, not everything can be drawn. Other fanciful illustrative techniques might be used, such as drawing a bridge where a road crosses a stream, placing larger-than-life boats in a harbor, exaggerating the size of a local ski peak, and so on.

See also: Flourish, Sketch

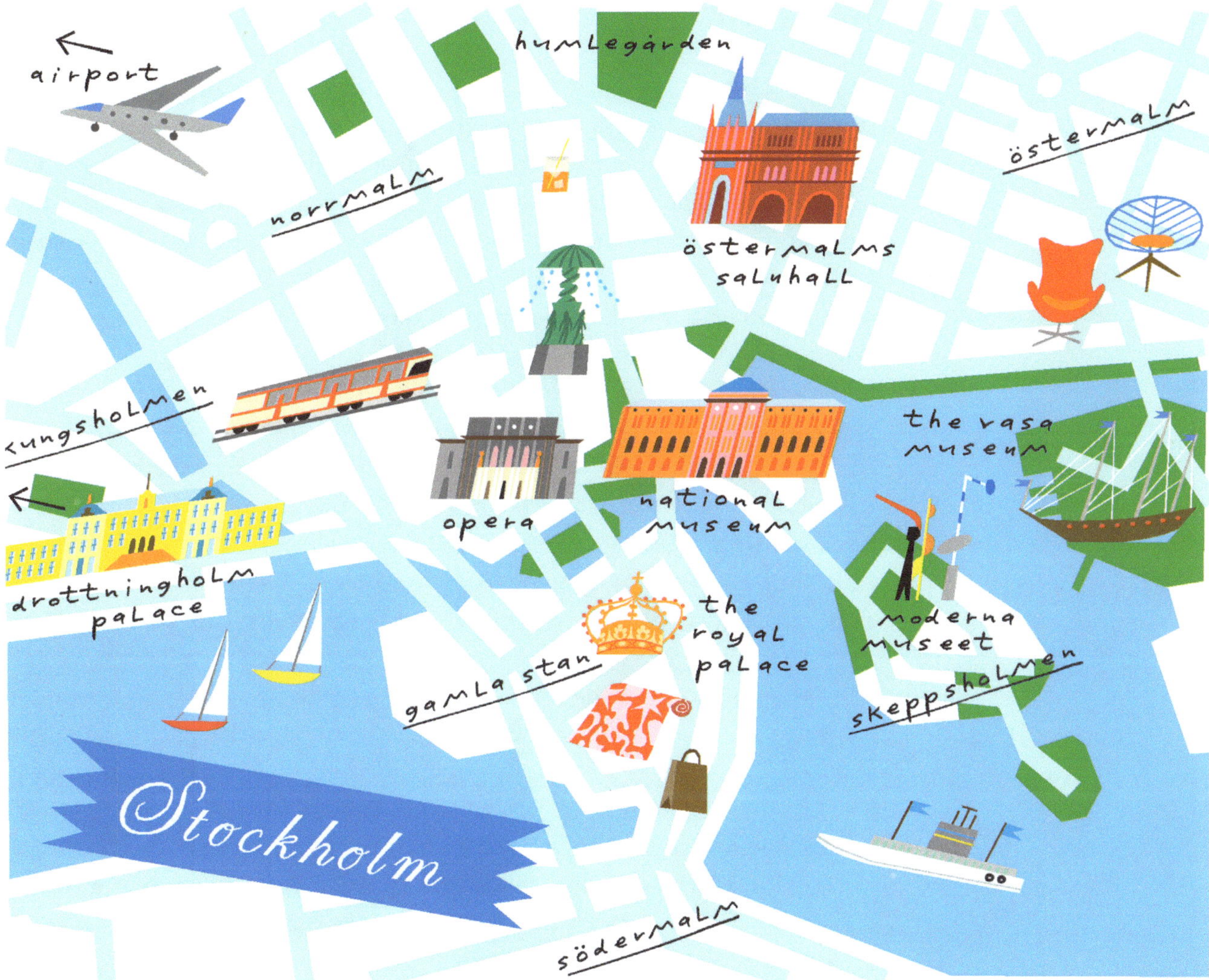

Stockholm by Lena Corwin.

Abstraction

USAGE Abstraction in maps is found in many forms, including atypical colors, repetition, purposeful feature distortion, unexpected labeling, and more. Sometimes an abstraction is just a small part of the map composition, while, at other times, the entire map is an abstraction. A humorous trend in mapping has been to label certain well-known features in extraordinary ways. For example, recently a U.S. map was created where the states were labeled not with their state names, but with "the best TV series" representing each state.[1] Often, the entire map is an abstract work of art meant to entertain (see previous example), make a statement[2], or provide visual interest.

STYLE While it would be difficult to describe a common style among maps with abstraction techniques, the one thing that abstractions have in common is that, while they may resemble real thematic or transportation-oriented maps, they are not intended to seriously be used for gleaning facts or navigation. Many abstract maps are artist's pieces and are thus created out of non-traditional mapping media such as paints, metals, clay, and even gelatin.[3]

TECHNIQUE Many abstract maps are riffs on existing cartographic products, whether abstract or not. While the ideas may seem novel, they are usually derivative of an inspiration piece. The example shown here was inspired by Andy Warhol's Marilyn Monroe series. In fact, the colors are near duplicates of those found in a few of the Marilyn prints, though the underlying data is a real streetscape of the Santa Barbara region.

Abstract maps are fundamentally creative endeavors. While all maps require the cartographer to enlist creative processes, these require it in full force. To create an abstract map from scratch, without an existing idea, creative exercises can be employed to gain insight. Exercises can range from 30-second doodles to several-hour brainstorming sessions. Consult the creativity literature to learn what exercises to do and how they will help.[4]

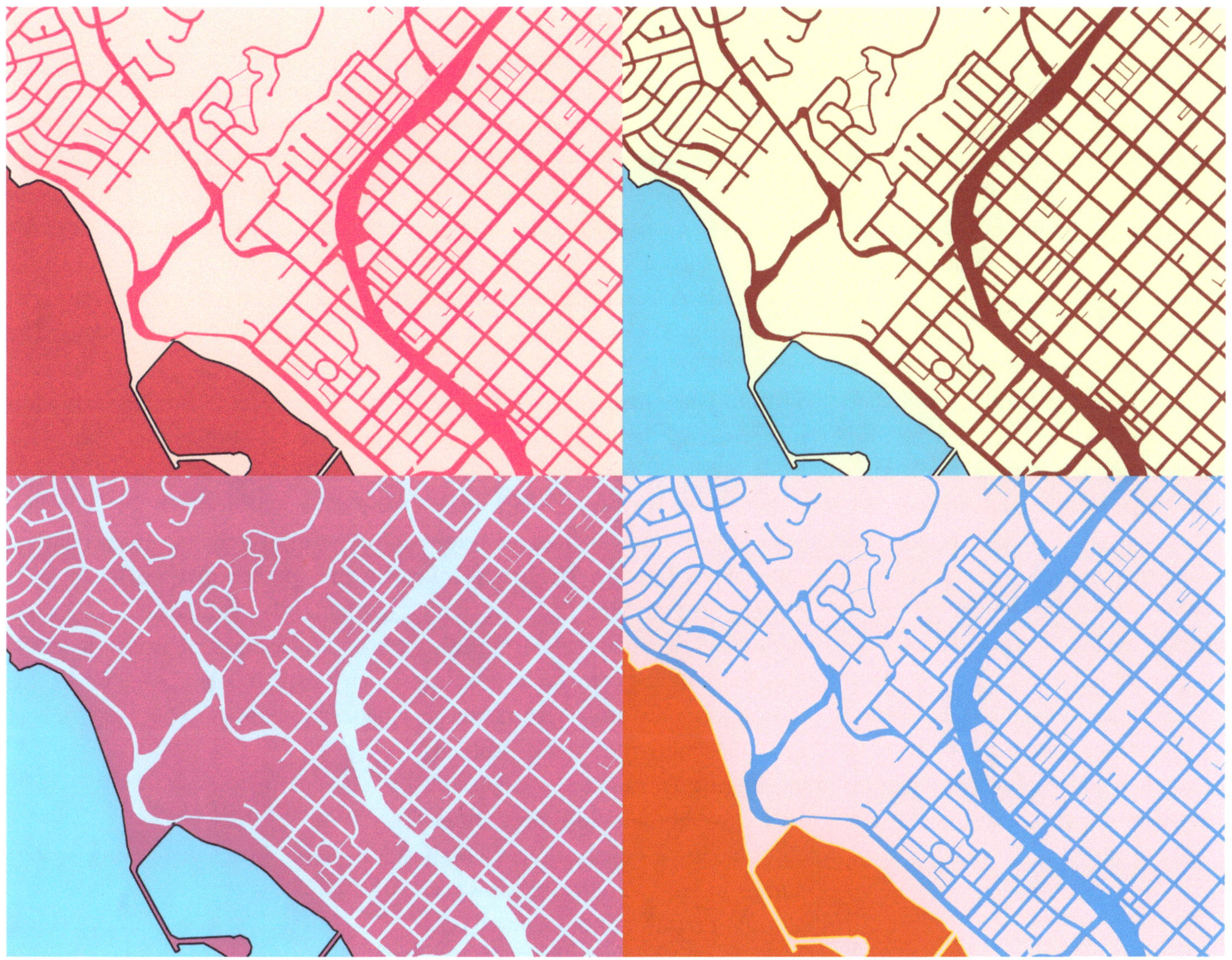

Abstract Santa Barbara by Gretchen N. Peterson, PetersonGIS.

[1]See "50 States, 50 Television Series", a map by Andrew Shears.

[2]See "Tight Spot", a giant inflated globe installation, by David Byrne.

[3]See the Jell-O map of the United States, commissioned by Kraft.

[4]One excellent creativity book is *Creativity Today.* Vullings, Ramon, Godelieve Spaas, and Igor Byttebier. *Creativity Today.* Amsterdam: BIS Publishers, 2007.

Alternative Borders

USAGE Very recently, there has been a surge in interest in making global-scale maps that display an alternative to standard political borders. Instead of the usual country boundaries, a map with alternative borders displays borders that are driven entirely by data. While some country boundaries will often appear in that data, others will disappear, and still others will show up in unexpected places. It is this comparison with real political borders that makes these maps so fascinating.

The datasets used to create these maps are massive, often gleaned from online sources that collect this information. Maps of these datasets can be created to show languages, family connections, software usage, university attendance, and so on. One typical example is a map of twitter languages that shows where, and in what language, tweets are being written. Some political borders, such as the Mexican-American border, show up distinctly in this map, while others, that share common languages, disappear. Large datasets with a human component and location information are required for alternative borders maps.

STYLE These maps, much like those comprised primarily of flow arcs, tend to be produced in an *Earth at night* style—which is to say, a black background with dots, or lines, of bright colors layered on top. The bright colors represent the data. This style trend allows for a clear and eye-catching display. More muted color styles are also satisfactory, and may become more so if the bold-color trend proceeds to ubiquity. In fact, a few maps with alternative borders do show a different style. Some—such as racial distribution in large cities—are shown as groups of multitudinous colored dots over a blank background.

TECHNIQUE A willingness to work with massive datasets—not to mention acumen—is required. These maps can be created out of many different datasets, as long as the data are dense enough to form quasi-political borders (or landmasses) and meaningful enough to provide new insights. The data must be at least partially human-centric in order to form at least a few recognizable borders. For example, a map of bird migratory routes done in the same style does not form alternative political borders, and depending on the species, may not even form recognizable landmasses. While there may be a compelling reason to create such a map, it does not fall under this pattern category.

See also: Flow Arcs

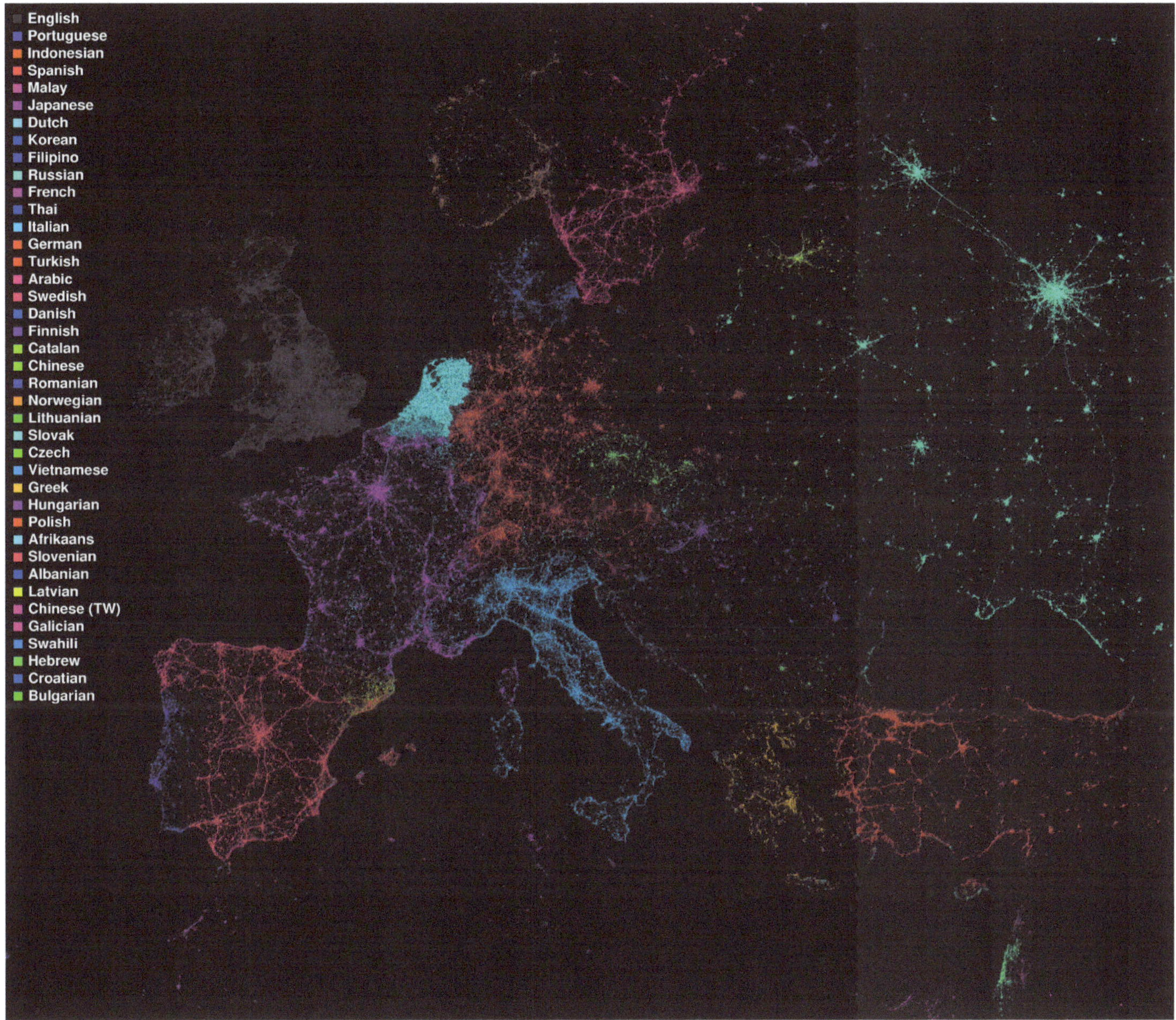

Language Communities of Twitter by Eric Fischer.

Micromaps

USAGE Micromaps are a recent addition to the cartographer's (and statistician's) toolkit, having been invented in the mid 1990s. As with other mapping devices, micromaps are used to illustrate the spatial dimension of a statistical data set, and are frequently shown side-by-side with graphs of the data. The maps, which are small in order to provide a general geographic overview of the statistics, are arranged in an organized way to enhance the data display. There are three types of micromaps: linked, conditioned, and comparative.

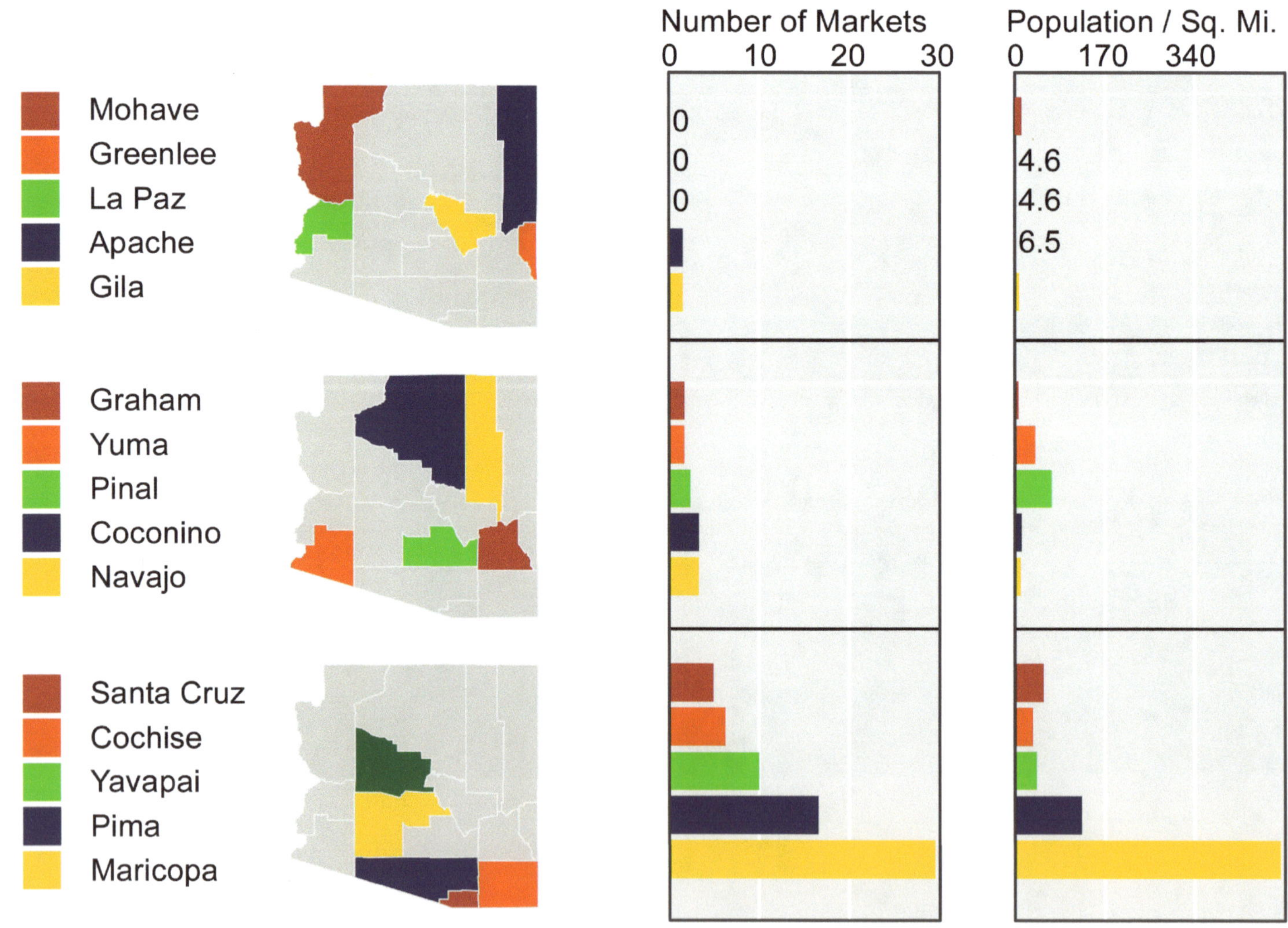

Arizona Farmers' Markets by Gretchen N. Peterson, PetersonGIS.

STYLE In linked micromaps, the colors used in the map match the colors used in the accompanying statistical tables and graphs, without regard to the magnitude of the variable(s). Linking the statistics by color allows for the display of bivariate and multivariate data, unlike in choropleth maps. This is accomplished by laying out the display in a series of rows and columns. The rows represent each region (e.g., states) and are labeled as such, and one or more columns show (1) the measurement and (2) maps of the regions. The measurements are denoted by colored dots corresponding to the region colors used in the maps. In this way, an additional dataset can be displayed by simply including a new column. As the map reader looks down the page from region to region, the regions are grouped (groups of 5 are shown to work well) so that only 5 regions are displayed on any one micromap.[1] Thus, the data can be visually analyzed horizontally or vertically to derive multiple conclusions.

Conditioned micromaps are similar to choropleth maps in that the region colors are used to denote the magnitude of the variable. In a set with many regions (e.g., U.S. States), the regions are grouped so that each micromap displays only a subset. These micromaps are then organized into a grid that helps interpret the data further. In other words, both the micromaps themselves and the organization of the micromaps on the page inform the map reader in conditioned micromaps.

Comparative micromaps are also arranged in an organized grid so that comparisons can be made among the maps, but instead of showing where each region falls in a choropleth style, they represent a time series where only regions that change significantly (e.g., move from one color in the classification scheme to another) are highlighted.

In some, but not all, cases, the geographic regions in the micromaps are drawn out of proportion when the regions vary greatly in size. This may happen when the smallest regions are not adequately visible due to the small sizes of the micromaps. Some micromaps of the U.S., for example, show the small states in the Northeast as larger than normal and the larger states in the West as smaller than normal.

TECHNIQUE When micromaps are composed of non-scaled regions, as described in the Style section, their strange appearance requires an explanation to the map reader. A caveat such as, "geographic regions are not shown to scale" is an appropriate accompaniment. Since using non-scaled regions on maps is not highly desirable, do not create them unless it is absolutely necessary.

In all cases, to create features that are visible at small sizes, a simplification algorithm (found in mapping or graphics programs) can be applied to the regions. This doesn't change the sizes of the shapes a great deal, but it does straighten the lines so that small curves do not fill in with ink when exported at the small size needed for micromapping. While some micromap regions are simplified almost to the point of caricature, it is advisable to try many different levels of simplification until the best compromise between feature visibility and abstraction is attained.

Comparative micromaps, where only the regions that changed to a different magnitude category are highlighted, take some trial and error to create a display that will be the least confusing but also the most informative. Because the technique of highlighting only the changed regions is uncommon in maps, the map reader will need to be given enough information and time to interpret the map accurately.

See also: Small Multiples

[1]Carr, Daniel B. and Linda Williams Pickle. *Visualizing Data Patterns with Micromaps.* New York: Chapman and Hall/CRC, 2010.

Small Multiples

USAGE A map composition containing a small multiples structure displays three or more maps in unison to enable rapid comparisons and analytical inferences by the map reader.[1] The small multiples pattern is typically employed when multivariate data needs to be shown for the same location (i.e., one small map for each dataset) or when a single dataset needs to be shown for varying locations.

It should be noted that interactive and animated map techniques are also possibilities for mapping multivariate or multi-location data, though these options do not allow the map reader to easily return to previous maps for on-the-fly comparisons. That is the strength of the small multiples strategy: the ability to quickly look at all the maps to more easily discern differences and similarities and thereby reach advanced conclusions. Maps using the small multiples pattern are used in syndromic surveillance, voting trend mapping, pollution type and location visualization, and weather trends, to name a few. This pattern is applicable to a large number of fields but is underutilized.

STYLE Small multiples maps are often laid out in a grid, and read from left to right. For small multiples of the same location, there might be one large map of the location that shows the location details (e.g., topography, city labels, and country boundaries), while the other maps are quite small, with minimal location information (e.g., landmass boundaries) and just the single dataset that needs to be displayed. Small, uncluttered maps facilitate quick eye movements and correspondingly fast comprehension rates. Small multiples of the same dataset in different locations contain more background information but are consistent in their symbology throughout. In this situation, only one legend is needed.

Most small multiples layouts lack color vibrancy, and in fact, many are completely grayscale. With many maps to look at, the choice to use grayscale is an effort to not overwhelm the map reader. Another style device is that each individual map is usually marked with a one -or two-word label denoting the location or name of the dataset.

In some small multiples designs, the background information is completely omitted, showing only the top layer data and how it changes over time. The idea is that if the map reader is already knowledgeable about the area, there is no reason to show it. For example, picture a few graduated circles that change in position or size in each map. This becomes very abstract, but if the map reader doesn't need to know the exact position of the features, only their general relationships, then it makes sense to test this method for ease of interpretation.

TECHNIQUE When displaying many maps of the same region, it is imperative that the extent of each map be *exactly* the same, since any small difference in extents will be distracting. Similarly, color schemes must be consistent. If each dataset in a multivariate display illustrates a heat map or choropleth, the color schemes ought to be the same so that, for example, red always means the most or perhaps "best" amount of a variable. When the maps show different qualitative datasets, the color schemes may be different, though the map reader will need extra time to interpret the map. Additionally, each map will require a separate legend.

To create the small, uncluttered maps that are required with small multiples displays, the cartographer should seek to minimize anything that will distract. Only a small amount of white space—or none at all—is needed around the individual maps. Like-

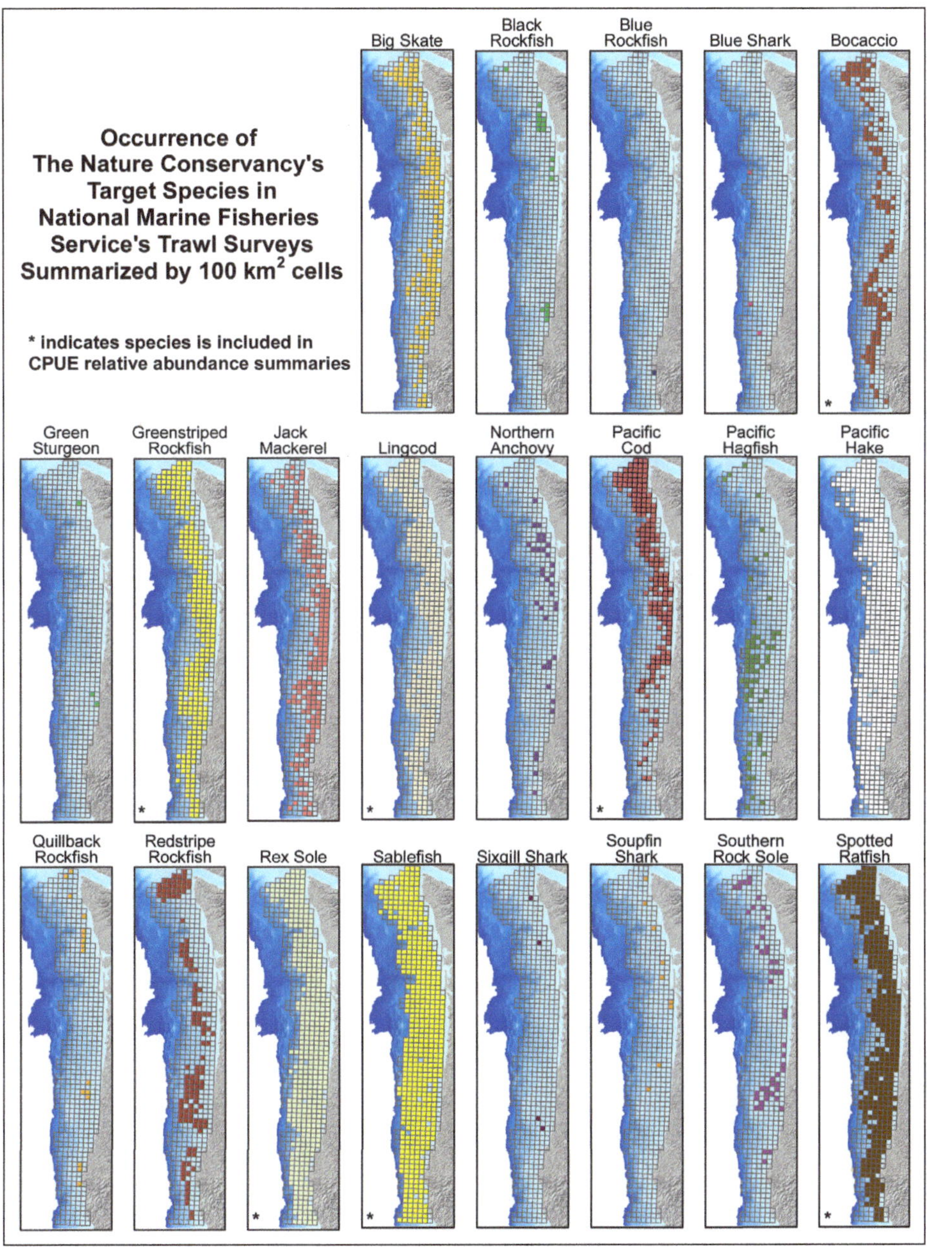

Target Species Occurrence by Allison Bailey, Sound GIS.

wise, do not place frames or borders around the maps, or, if a small amount of separation is necessary, a thin, simple line should be all that is needed.

See also: Animation, Interactive

[1]See *Envisioning Information* for a broad discussion on small multiples, including non-map small multiples. Tufte, Edward R. *Envisioning Information.* Cheshire, CT: Graphics Press, 1990.

Information Graphics

USAGE Information graphics, or infographics, are data stories that present a dataset in a transparent and effective manner in order to explain a phenomenon or to persuade an audience. The data is usually complex, but presented clearly. While it could be argued that information graphics have been around for centuries, the modern take on them started as recently as the 2000s and has since risen in popularity. They are most often found in media such as newspapers and magazines, though they are expanding to be included in just about everything from publications, to posters, to websites.

This section focuses on information graphics that include one or more maps (not all information graphics have maps). These infographics dis-

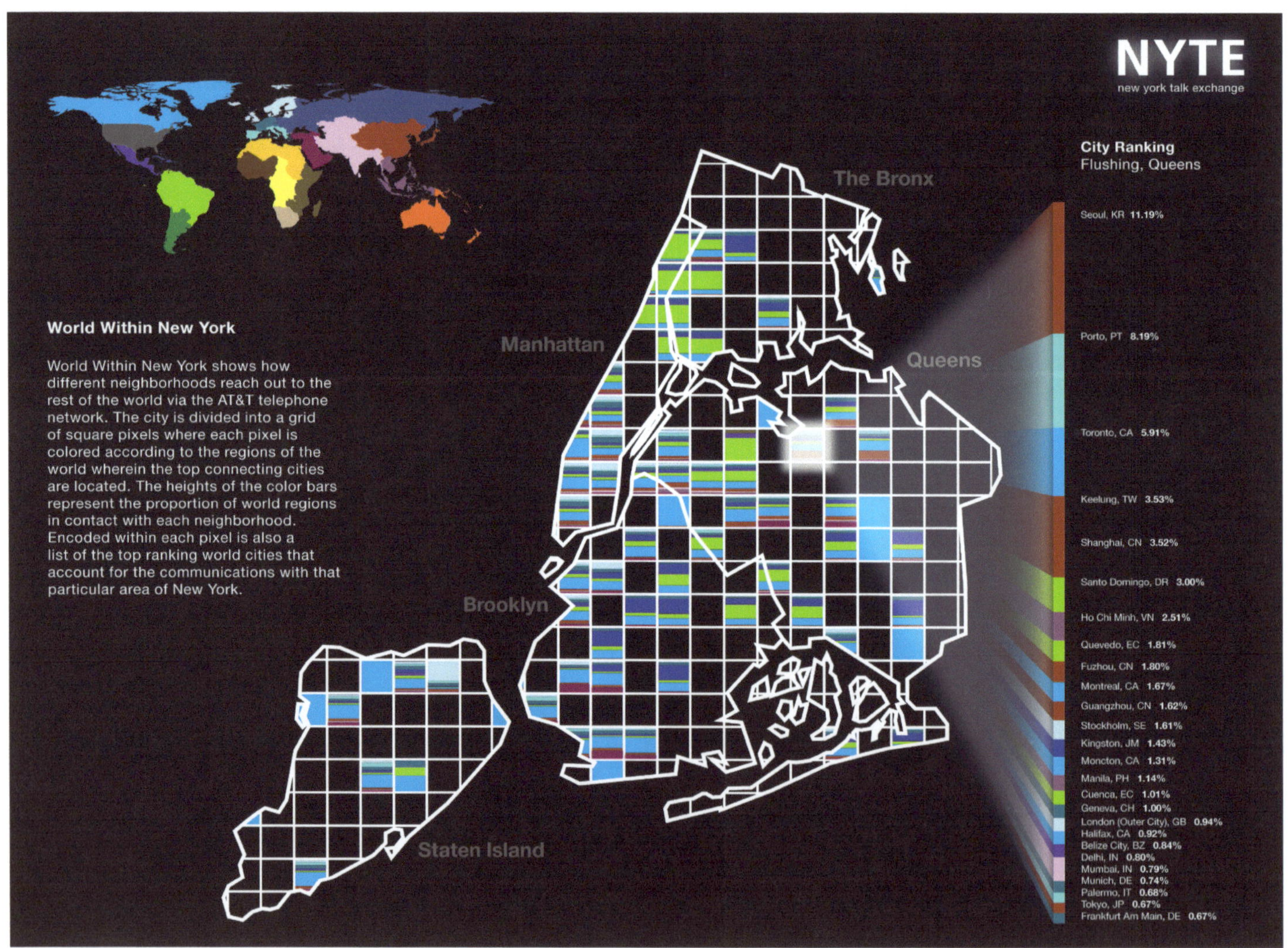

New York Talk Exchange. A project by the MIT Senseable City Lab senseable.mit.edu/nyte/

play a compelling conclusion garnered from a dataset in a way that engages the reader to explore the results spatially (and often temporally) as well as through explanatory text, graphs, pictures, charts, and any other supporting material. The reader engagement that comes from the interplay of information on the page establishes trust in the information and analytical results, stemming from the dual purposes of presenting all the information the reader needs, while maintaining clarity and, sometimes, sequential flow.

Most information graphics are in the form of static poster-style displays, whether printed or digital. More recent contributions to this genre include animated maps that, instead of being explained via the usual static texts and graphs, are explained via audio. These are, essentially, movies of maps. Movies of animated maps accompanied by audio are created to teach historical movements of people, political boundaries, and armies, to name a few.

STYLE Maps on information graphics need to be well integrated into the surrounding page by using the same colors, line-styles, and typography as the rest of the page. In other words, maps should not appear as though they were created in a separate program and then copied onto the page. Any elements, such as region boxes, graduated circles, or points that are layered on top of the basemap are also in the same style as the rest of the map. These top-layer elements will sometimes lead off the map itself and onto the rest of the page layout. For example, leader lines, a series of bubbles, and call-out boxes are all possibilities.

Map styles in information graphics are some of the most innovative in cartographic design today. Infographic maps often have very simple basemaps—gray landmasses with white country borders, for example. In order to easily show large quantities of information, they may have leader lines emanating from the centerpoint of each country or state that connect to a graphic that displays quantity. In such displays, the map reader can glean information from the map and the graphic separately, and even gain a deeper understanding when exploring both in tandem. This provides an alternative to choropleth mapping.

TECHNIQUE In most cases, it is best not to compartmentalize the map, which means that placing a frame or border around the map is counterproductive to the cohesive look that is desired. The maps are often quite simple, with minimal background information, in order not to compete with the other elements on the page, while still supporting the presentation. Graphic design elements such as special insets of certain locations and arrangement of the map around the other page elements are expected.

Because the storytelling capability of the information graphic is of the highest importance, all maps created for these displays must also have that storytelling ability as their ultimate goal. Flourishes and any other non-useful items must be minimized. The infographic itself may have flourishes if it fits with the theme, but the map generally does not.[1] In fact, information graphics are usually created entirely within graphics programs because the map elements are simplistic enough to do so.

[1]See the infographic titled "Best Beer in America" by Mike Werth for an example of an infographic with some flourish.

Thematic 3D

USAGE Thematic 3D mapping refers to the extrusion of features or symbols in proportion to the magnitude of a variable. (A thematic 3D map is also called a prism map, which itself is a form of choropleth map.) This pattern is useful when the map reader needs to see general trends in data or when flashy graphics are required. The cartographer is warned that this pattern often results in rudimentary-looking maps and artistic bias; choose this pattern only when no better solution presents itself or when a lack of scientific rigor is acceptable.

Thematic 3D maps are displayed on report covers, on website introductory screens, in presentation slides, and anywhere that intense scrutiny is unanticipated. The subject matter for thematic 3D maps has included the

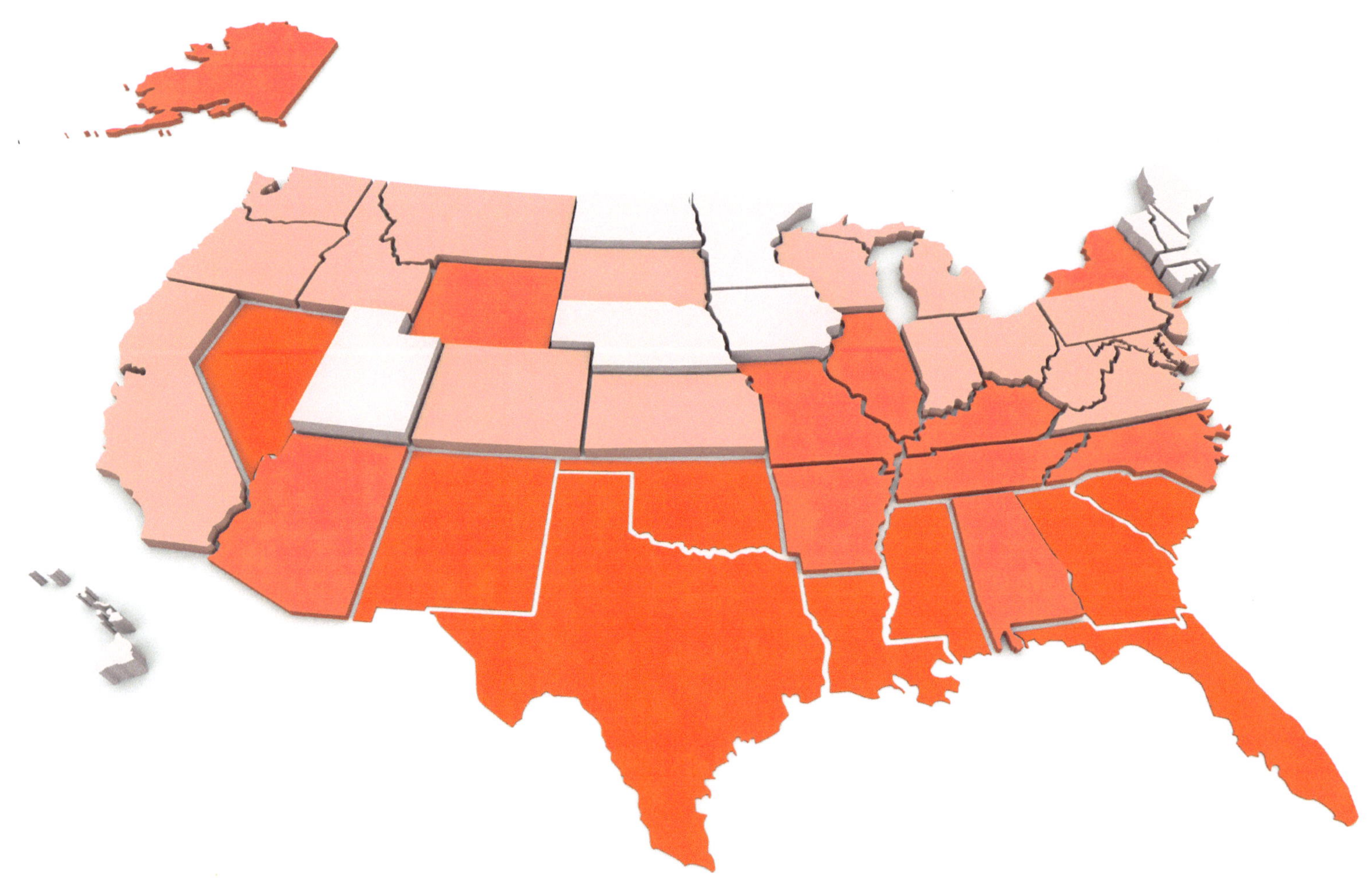

Health by State by Chris Lamphear.

mapping of population, religion, immigration, age, housing density, and more. They have even "popped up" in art galleries, in true 3D.

STYLE Thematic 3D maps are commonly small-scale representations of entire countries or the whole globe, though they do not have to be so. A familiar style is a population density map where the density is represented by bars of varying height. In some maps, the basic bar style is further enhanced by using denser data (not just major cities but all population centers, for example) and thinner bars or lines that create a spike effect in the most populated areas.

Other maps with this pattern show country or state polygons—the shapes of the features themselves—extruded by any number of variables pertaining to those areas. Greatly exaggerating the extrusion can be somewhat alarming, though that may, in fact, be intentional. Linear features have been mapped in 3D—for example, representing traffic as a "wall" that varies in height according to vehicle density. Gridded data can also be mapped in this way, so that the grid cells appear to be "pushed out" to varying degrees and potentially color coded as well. This gives an appearance similar to a pin impression toy.

A thematic 3D map typically does not have many data layers besides a simple background and the 3D features. Labels on the highest and lowest points, often connected by leader lines, are used when this additional information needs to be highlighted. In order to create a meaningful result, the height proportion may be inconsistent. For example, the highest values might be four times the height of the smallest values even if they only represent a doubling in magnitude. This is done to emphasize the difference between high and low values, but can mislead the map reader.

TECHNIQUE In situations where the variable range for particular features needs to be conveyed to the map reader, a graduated color scheme needs to accompany the 3D extrusion in order to facilitate legend lookup. This is because it is difficult to precisely match up heights with a map key. Once the color scheme is applied, however, the cartographer must then decide whether or not the 3D effect is necessary.

Most mapping programs have the capability to create 3D symbols on the center points of polygons or on points themselves. Some thought as to what symbols are used is warranted. Usually, these symbols take the form of bars or, sometimes, thin lines, as mentioned in Usage. Since thematic 3D effects are often used for general trends and flashy graphics, consider using symbols that literally represent the data. For example, a thematic 3D map showing cow density across the U.S. might use varying-height 3D cow symbols.

Great leeway is left to the cartographer in deciding how much to extrude certain values and whether or not the extrusion itself remains constant or is skewed to further highlight large or small values. As a result, the thematic 3D pattern is more of an art than a science and can therefore sometimes bring undesirable bias into the result.

See also: Representative 3D

Representative 3D

USAGE Representative 3D mapping depicts map features in a realistic way by extruding topography, buildings, and other features with an inherent height component according to their true, or relative, height, and in doing so, aids the map reader's understanding of the nature of the region. This is useful for attention-grabbing brochures and presentation slides, of course, but aside from wowing an audience, representative 3D mapping is particularly effective any time the height component is essential to the thing being mapped (e.g., pollution plumes, soil horizons, view sheds). A more recent capability in this realm is georeferenced 3D video, which can be played as layers on an interactive map, though this use is not yet fully explored. Note that a plan-view map, even if it contains hillshading or contours, is not considered a 3D map.

STYLE An oblique view is usually applied to the map. Stylistically, the same principles that apply to good 2D map design also apply to good 3D map design. The two types of oblique maps commonly seen are perspective (central) and parallel. Perspective mimics human sight and is therefore fairly easy to comprehend. Perspective also has several drawbacks—the most important of which, from a cartography standpoint, is its lack of consistent scale. In perspective, the lines of projection converge, making far-away objects appear smaller than close-up objects. Parallel oblique projections do maintain a consistent scale throughout the map image, allowing for accurate measurements. In a parallel oblique projection, scale is preserved because the lines of projection are parallel.

A representative 3D map can be realistic, cartoonish, or somewhere in between. Most are rendered digitally, but a few extra-special hand-painted specimens are to be found in niches such as ski trail mapping. Mountains and buildings are the most commonly depicted features on these types of maps. Less common, but just as interesting, are 3D forests (natural or commercial), oceanscapes, and true 3D models of towns, cities, arboretums, and national parks.

TECHNIQUE The viewing angle of the map needs to be set so that important features are not obscured behind back-facing slopes or buildings. If the map contains a focal point to emphasize, this area should be brought as much toward the center, or just off-center, as possible. Realism is applied to features in these maps via pseudo natural mapping (particularly for natural features) or by using building textures, which are specifically created in 3D design packages to produce realistic building facades. Building facades and other textural techniques are not always needed, however. A viewshed map might need just the basic building shapes, depending on the complexity of the analysis and the amount of clutter the building textures would contribute.

See also: Pseudo Natural

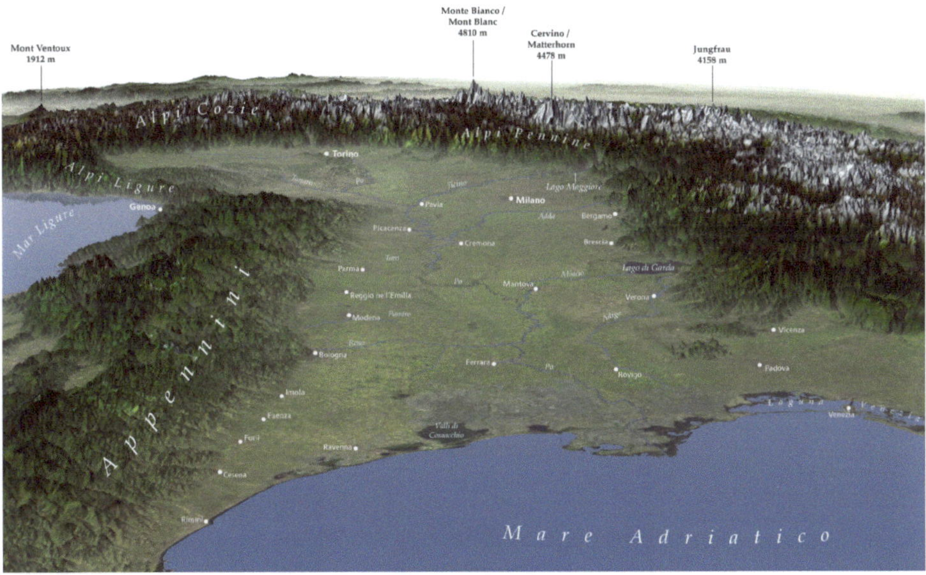

The Po Valley by Hans van der Maarel, Red Geographics.

Diagrammatic

USAGE Probably the most distinguishing feature of a diagrammatic (schematic) map is a lack of consistent scale, while maintaining, for the most part, accurate feature chronology. In other words, parts of features may be stretched out of proportion, but they still line up in the order in which they appear in the real world. The most common application of this map style is in mass transit mapping. In these, inner-city locations are stretched to a further degree than the outlying locations in order to fit everything onto one page; this is done for portability and to provide extra space for labeling the much higher numbers of paths and stops located in the inner-city. Less common diagrammatic applications include solar system maps, brain maps, and even cartograms, all of which stretch or shrink the features—usually countries—by the magnitude of a variable. The focus in this section is on transit and cartogram styles.

STYLE Diagrammatic maps are easily identified by their visual similarity to infrastructure schematics (transit maps), by their warped feature shapes (cartograms), and by their lack of true scale (both). Background information is usually kept to a minimum. Transit maps are often very complex even without any background information, and cartograms may be difficult—sometimes impossible—to interpret, even without interference from background noise. Both styles contain simplified line work to a much greater extent than on other maps.

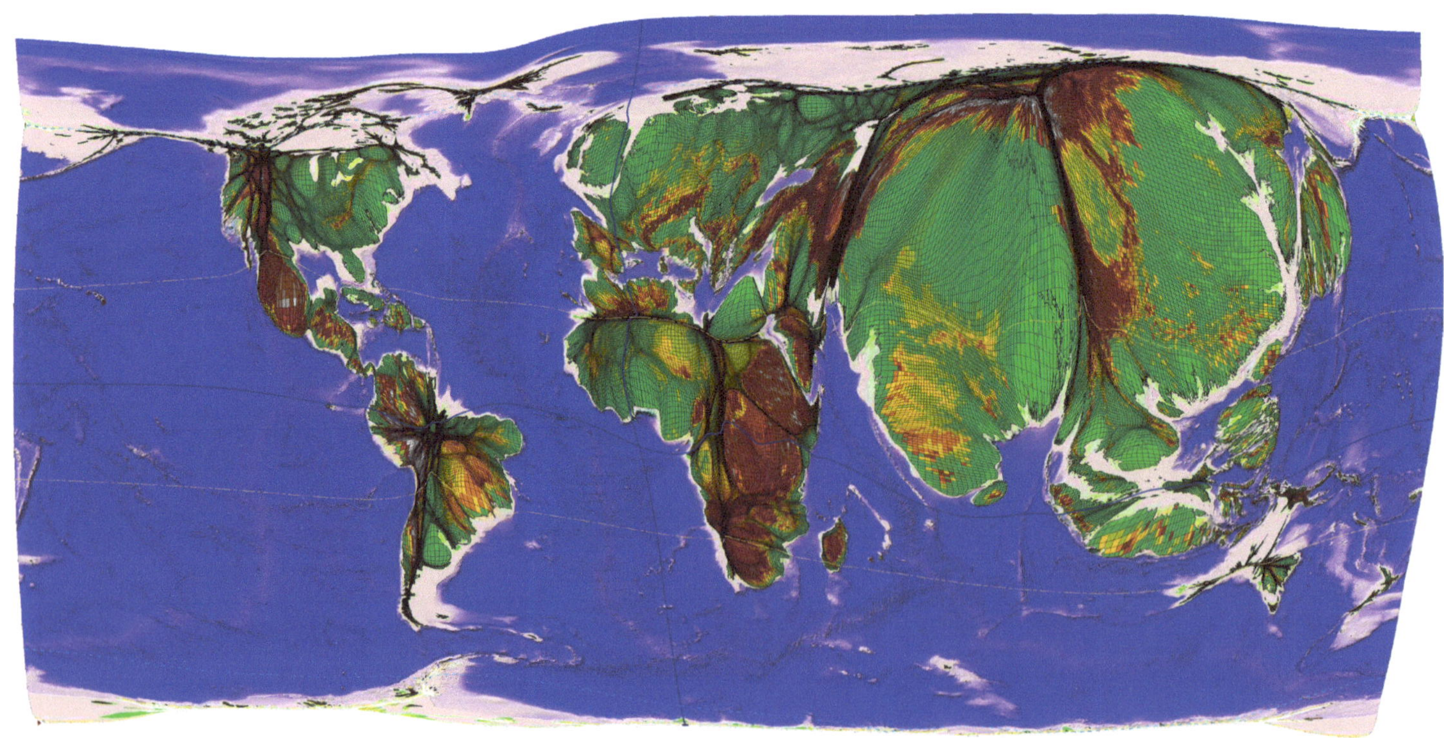

World Topography displayed on Gridded Population Cartogram by Benjamin D. Hennig.

TECHNIQUE The geographic locations of features are not as important in diagrammatic mapping as in other maps, but the relative positions of features must nevertheless remain intact. To produce a diagrammatic map from geographically accurate data, it is best to begin in a mapping program and export to a graphics program to do the bulk of the design work. In transit maps, labels and stops will need to be placed individually, with one eye on ideal placement positions and the other eye on relative accuracy. For example, if the first three bus stops are within one mile of each other but the fourth is two miles from the third, the first three stops and labels can be placed nearer to one another than they are to the fourth, if space permits.

In cartograms, the simplification of line work and distortion of shapes is generally (but not always, some are hand-drawn) achieved via a specialized cartogram algorithm.[1] In transit maps, it would be possible to begin by simplifying the transit lines with a simplification algorithm, but the results would likely not be adequate. For transit maps, then, it is best to begin with spatially-correct geographic data and proceed to modify it onscreen, or to simply draw new lines using the geographic data as a guide.

The map reader might benefit from having a companion map that shows the diagrammatic map superimposed on what's known as a *distortion grid* that is essentially a 3D mesh, with larger grid cells corresponding to places where the scale is stretched and smaller grid cells corresponding to places where the scale is scrunched. The distortion grid gives the map reader an immediate understanding of where and by how much the scale is distorted.

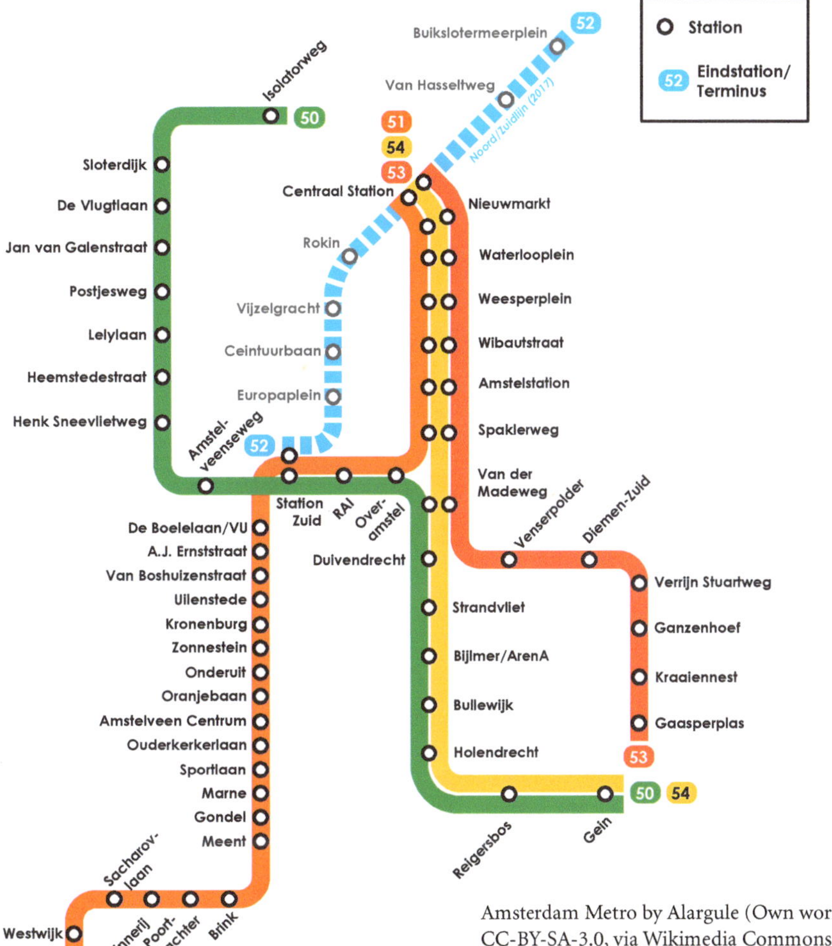

Amsterdam Metro by Alargule (Own work), CC-BY-SA-3.0, via Wikimedia Commons.

[1]See Scape Toad software developed by Dominique Andrieu, Christian Kaiser, and André Ourednik, for one example.

Choropleth Maps

USAGE A choropleth map represents a continuous variable via a color progression (graduated color scheme) within discrete features such as countries, states, or watersheds. For example, states with high infant mortality might be shown in red, while states with low infant mortality might be shown in yellow. The hue for each state can be matched with a legend that records the rate (or range of rates) associated with that hue. The choropleth technique is usually applied to area features but can also be applied to line and point features. Some limitations of the choropleth *pattern* are:

1. No inferences about the mapped variable can be made at a finer scale than that which is mapped. A map of body-mass index by state, for example, shows nothing about how that variable is distributed across counties within the states.
2. The changes in color between features can make it seem, to the novice map reader, that the variable changes abruptly at those points, when in reality the variable probably changes much more smoothly.
3. If the features being mapped have widely varying areas, the map reader may misinterpret the larger areas as being more important, even if they have the same value as smaller features.
4. Bright colors can be interpreted as being more important, even if this was not the cartographer's intention.

STYLE Usually, background data and other map details are kept to a minimum on choropleth maps, as they tend to clutter the display. Bivariate choropleths—where there is one color scheme for one variable and another color scheme for another variable—are not combined on the same map in a static presentation. In a static map presentation, bivariate choropleths are placed on individual maps, often as small multiples. Interactive maps usually present bivariate choropleths on the same background map but with a toggle for the user to turn each choropleth on and off, with no two being displayed at the same time.

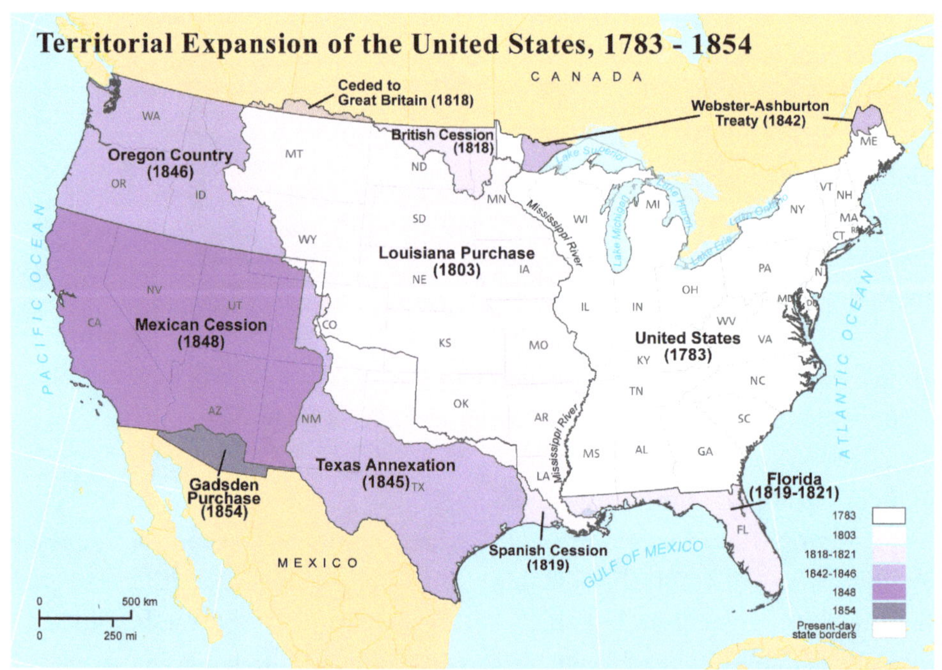

U.S. Territorial Acquisition by Gretchen N. Peterson, PetersonGIS.
Reprinted by permission of DWJ BOOKS and CQ Press.

TECHNIQUE Choropleths are easy to create in most mapping software. The major cartographic consideration is how to break up the data

Method	What It Is	Not Good For	Good For
QUANTILES	Equal number of data values in each category	Skewed data	Comparisons across time or between different datasets
EQUAL INTERVALS	Equal data value ranges	Skewed data	Comparisons across time or between different datasets
JENKS	Minimum variation within categories, maximum variation between categories	Comparisons across time or between different datasets	Most data including skewed data
NESTED MEANS	Categories above and below mean (2), above and below sub-means, etc.	Non-even numbers of categories; skewed data	Normally distributed data where mean is an accepted break-point
GEOMETRICAL INTERVAL	Breaks are based on a geometric series; variance is minimized within classes	Understanding by the Average Map Reader	Skewed data with many duplicate values.

into ranges that make sense, which depends on the numerical distribution of the data—especially when it contains outliers or a skewed distribution. Different strategies for breakpoints exist. To begin with, if there are set breakpoints that are established for the data, use those. For example, a map of body-mass index will have a breakpoint at 30, since this is the obesity threshold. However, many datasets are too unique to have commonly accepted breakpoints.

The cartographer also needs to consider normalizing the data if it is appropriate. For example, a dataset showing where schools are located in a state may need to be adjusted for population in order to obtain a meaningful map. The most common normalizing factors are population and area. Without normalizing the data, the map may simply show the underlying population or area trend rather than the trend in the focus data. If previously established breakpoints are not going to be used, and the normalization issue has been dealt with, various options for how to break up the data into appropriate bins will need to be considered. See the table on this page for more information on these classification methods.

The color progression can consist of a single hue that varies in brightness from low to high, or a scheme with multiple colors. Multicolor progressions use two different colors, one for the highest and one for the lowest class; each progresses toward a moderate color for the median class. The number of gradients in a single progression should be kept to five or fewer—beyond this, the map reader cannot make an accurate match between the legend and the map. Furthermore, the color progression between classes needs to remain consistent. A large change in color from one class to another can lead a map reader to the inaccurate conclusion that those classes are further apart in magnitude than they really are.

Heat Maps

USAGE A heat map represents a continuous variable through the use of a graduated color scheme; the color scheme, as its name implies, is usual thermal. As opposed to choropleth maps, heat maps depict the magnitude of a variable—density, usually—without regard to the underlying geometry and usually in a spatially continuous manner. In essence, the data is "splashed" across the map to show where there is a lot of something relative to where there is a little of something. Note that *heat map* is also used to signify a two-dimensional, non-spatial matrix with a graduated color scheme; this is not the way the term is used here.

Heat maps, and their progeny *hot spots*, can pinpoint areas that require interventions, such as crimes, infrastructure decay, or biodiversity loss. They can also pinpoint more positive areas, such as access to fresh food, walkable neighborhoods, or high employment.

STYLE As mentioned in Usage, the color schemes used in heat maps are most often of a thermal variety: a low amount of the variable in a light color and a high amount of the variable in a dark color, usually red. For hot spot rendering, the colors are applied to concentric rings that emanate out from the hot spot areas, where a continuous gradient of color—from dark in the central rings to light in the outermost rings—is applied. Typically, the heat map data is semi-transparent and layered onto a basic location basemap to provide a sense of where the high and low points occur relative to landmarks and roads.

TECHNIQUE Because the color scheme for heat maps is continuous, the legend of a heat map can represent that continuity as well. A continuous color scheme can be represented as discrete boxes of color in the legend but can also be shown with a color gradient along with labels for the minimum and maximum values. In some cases a legend may not be necessary, such as in a newspaper overview map of crime hot spots, if the colors are self-explanatory. With Western-centric map readers, red will be automatically interpreted to signify a higher magnitude of the variable than yellow or light green, for example. Though not common in general overview maps, hot spots can be annotated in the map with other details that are needed to interpret the visual. For example, a crime hot spot might be labeled with explanatory text concerning the particular crimes associated with that region.

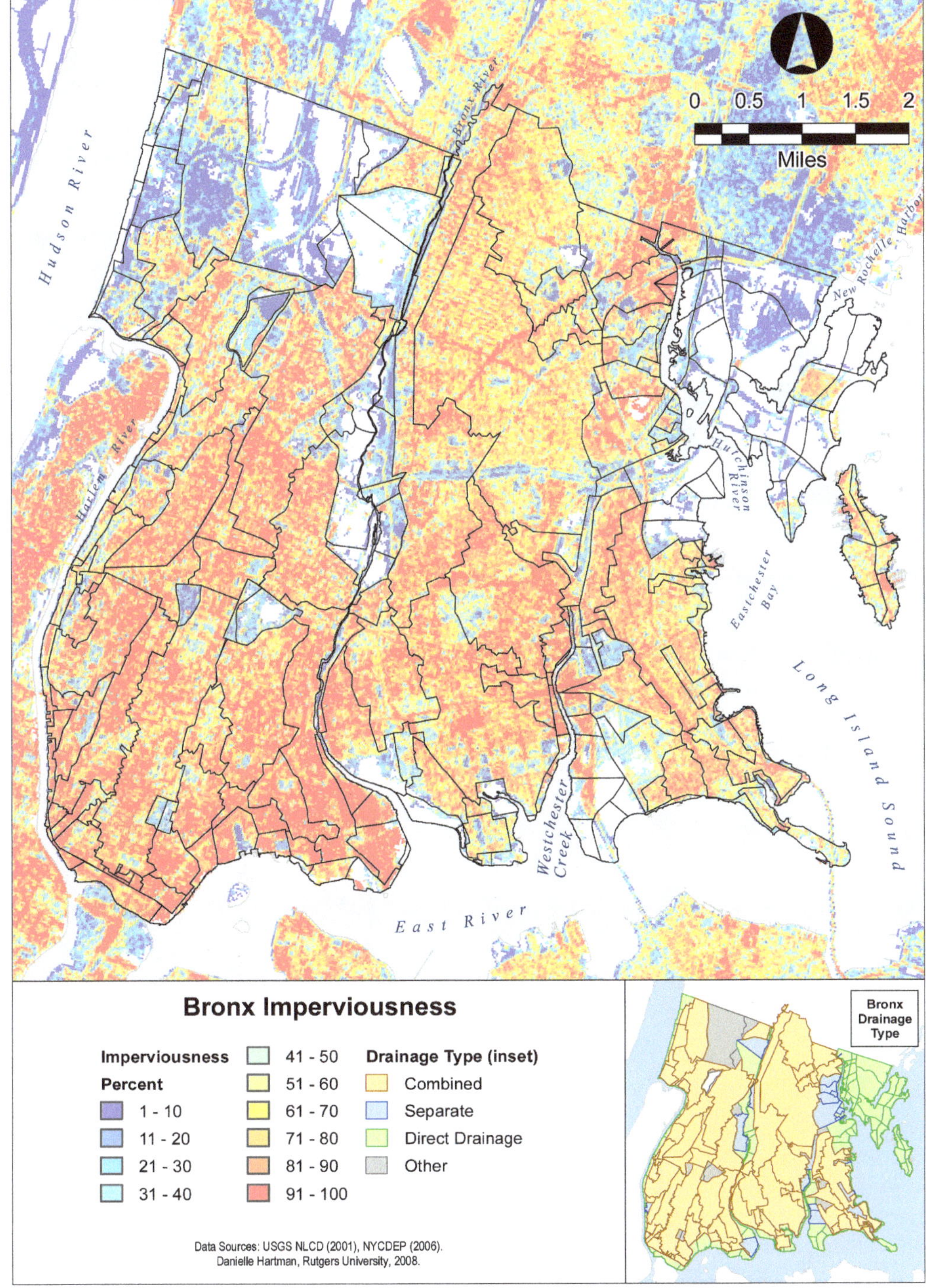

Bronx Imperviousness by Danielle Hartman, Rutgers University. © 2008 Danielle Hartman.

Type as Map

USAGE Type maps, or typography maps, have labels that mark spatial features rather than points, lines, or areas. In effect, the labels become the points, lines, and areas through label repetition. Current examples of these maps are almost exclusively artistic impressions of cityscapes, due to the density and linearity of features in those regions.[1] With some additional effort and creativity, they can be created for any type of region.

STYLE Most of the words on a type map are in uppercase to give a neater appearance, so that there are no ascenders or descenders to break up the visual pattern. The choice of typeface for each feature type is obviously important. Narrow character glyphs and bold weights are preferred because they create a denser appearance. Large blocks of repeating text can create unintended line or area-like artifacts, which is resolved by staggering the text on each line in a randomized manner. The style of these maps is impressionistic—the repetition of words provides a much different experience when viewed up close than when viewed at a distance.

TECHNIQUE Tedium, or, more euphemistically, *attention to detail* is the operative technique for these maps. Typically, one begins by blanketing the map page with a single word or phrase, where each word or phrase is separated by a character, such as a centered dot. Then, the smaller features (e.g., roads, buildings, or canals) are layered on top, thereby "cutting out" the underlying words. The top layers are further separated and delineated from the background by surrounding the text with halos or boxes that are the same color as the background. Each label intersection—such as road intersections—needs to be examined individually to determine which label should be shown; this decision may be driven by the relative importance of the road or the label that looks the best.

The choice of typeface for each type of feature is something to carefully consider. It goes without saying that the overall character of the map will be dependent on the personality of the typefaces more so than on standard maps. Beyond the typeface's personality, it is also important to choose varieties with bold weights and narrow character widths in order to achieve a pleasing hierarchy. These varieties of typefaces will also minimize as much white space around the glyphs as possible, which allows

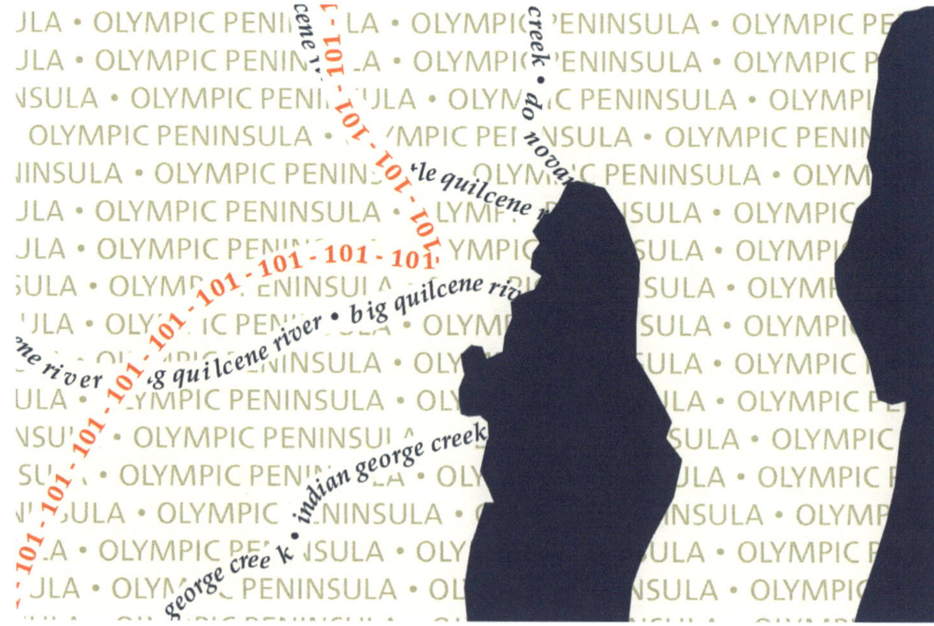

Olympic Peninsula by Gretchen N. Peterson, PetersonGIS.

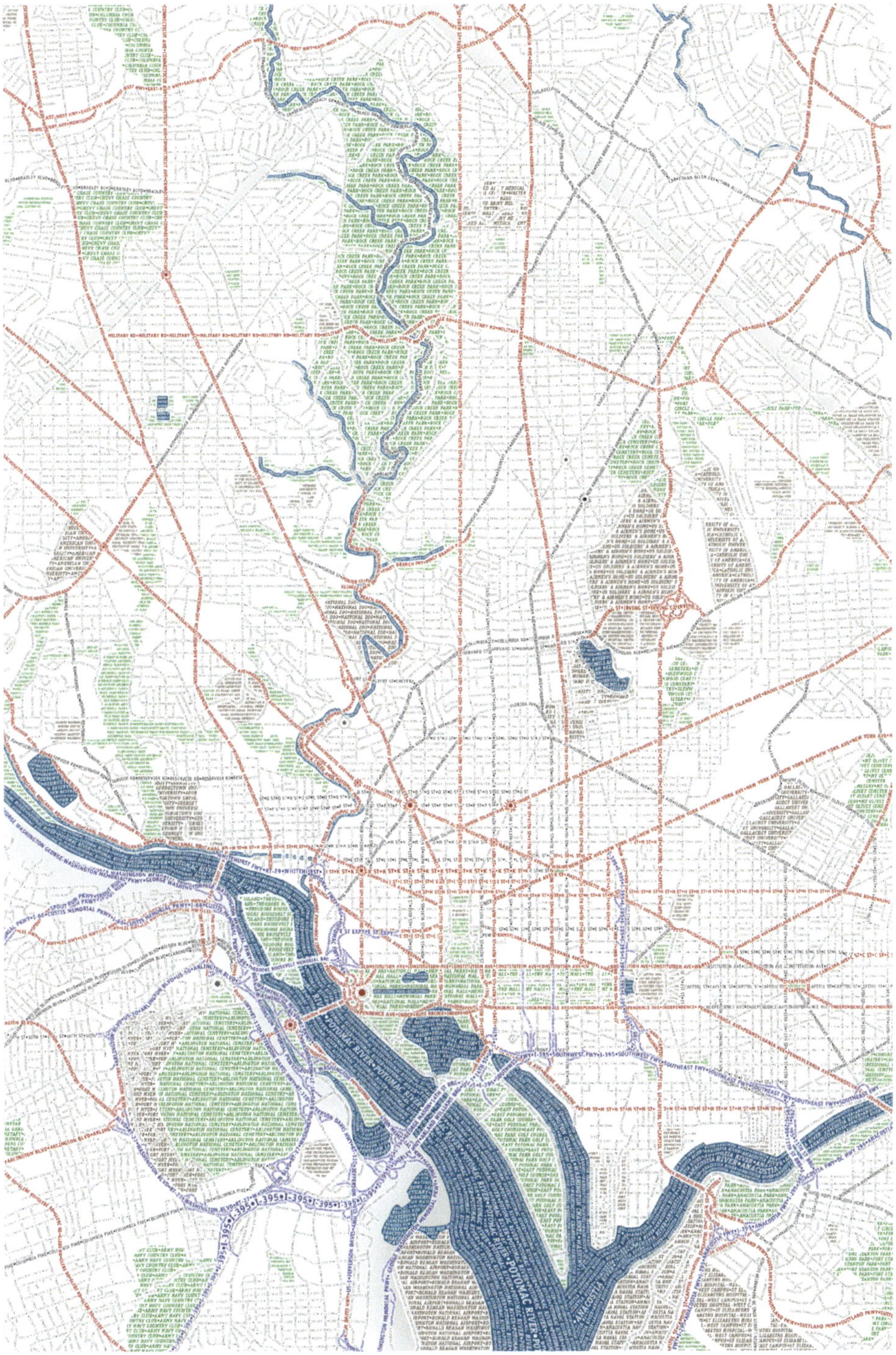

Washington D.C. by Axis Maps LLC

the map to appear more solid when viewed from a distance.

It may not appear at first glance that color will be an important consideration on type maps, but it is. Trying different color schemes and aiming for high contrast between the main blanketed word or phrase and the other words is important. To provide additional contrast, it may be necessary to consider a solid color fill under the words of smaller, linear features such as streams. While it is true that the cartographer of a type map should try to use *only* typography to define the areas on the map, using a solid fill in a few places may be necessary if enough contrast cannot be achieved via typography alone.

[1]The best-known typographic maps in the cartography world are by Axis Maps. Look up examples of Paula Scher's more free-form typographic maps for additional inspiration as well.

Polar Aspect

USAGE Polar aspect projections display a polar region in a realistic way. They are tangent at the poles; imagine your map as a flat piece of paper with its center point touching a globe at a pole. Polar views provide an alternative and more accurate depiction of the polar regions compared to the more common equatorial aspect maps, which often distort the poles to a great degree and are therefore difficult to interpret in the polar regions.

Polar aspect maps are seen in maps of arctic exploration, animal movement, political ownership dispute, climate, and more. However, they are one of the least frequently used projections types. Because the polar aspect map is a relatively rare projection type, it is included here as a *pattern*. The polar aspect's inclusion in the cartographer's toolkit assures that this useful projection-type won't be overlooked.

STYLE Many polar view maps are encapsulated by a bounding circle instead of a bounding box (frame) due to the natural circularity of the hemisphere geography from this point of view. (The example shown here has a rectangular bounding box, however.) In most cases, the cartographer restricts the mapped area to within a single hemisphere to minimize distortion. An artist's map may stretch the mapped area beyond the hemisphere, although spatial accuracy is then greatly compromised. Both hemispheres may be mapped on the same sheet more accurately by placing the two hemispheres side by side.

Depending on the projection chosen, the accuracy of area, distance, or angle can be maintained, with these varying in some projections as distance from the pole increases. On North Pole aspect maps, the International Date Line commonly runs toward the top, but the map can be rotated to any desirable configuration.

TECHNIQUE The main technique to be mindful of is choosing a suitable projection. In mapping software, the needed projections for this pattern are sometimes grouped into *polar projections*, though these are actually polar aspects of projections that can also be viewed in the equatorial or oblique aspects. The stereographic projection, centered at one of the poles, is popular for this type of mapping, but others, such as orthographic or gnomonic, may be used as well.

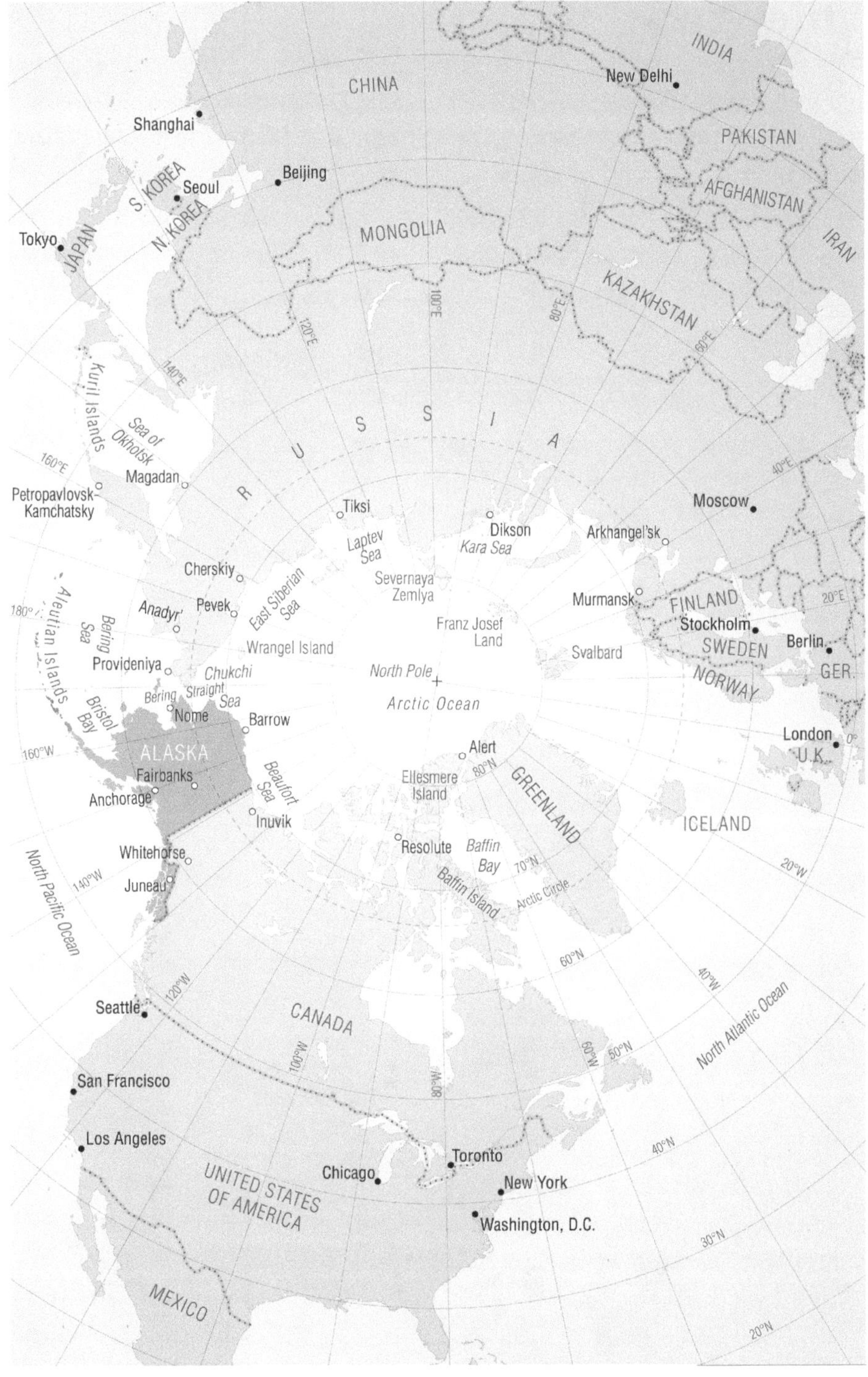

Polar View Map, North Pole by Nick Springer, Springer Cartographics LLC.

Latitude and longitude graticule lines should be added when possible, in order to aid comprehension of this unique aspect. They do not have to be numerous, and they can be thin, dashed, or dotted to minimize their appearance. As with graticules on any map, the best option is to run them underneath landmasses, though, occasionally, maps are created where the graticules run over the landmasses. When layering them under the landmasses, the cartographer might consider an artistic effect whereby the lines end just before they intersect with the landmasses.

Flow Arcs

USAGE Maps with flow arcs, also known as network flow maps, illustrate connections between points. Often these connections represent the flow of something—air travel, diseases, goods, information—but they can also represent distances, personal relationships, and other more abstract datasets. Direction and magnitude of flow are also represented on some flow arcs. Flow arcs do not need to comprise the entirety of the composition. For example, an otherwise standard map can include just a few arcs with arrows showing the direction of migration of a bird species. Conversely, an entire map can be comprised solely of flow arcs.

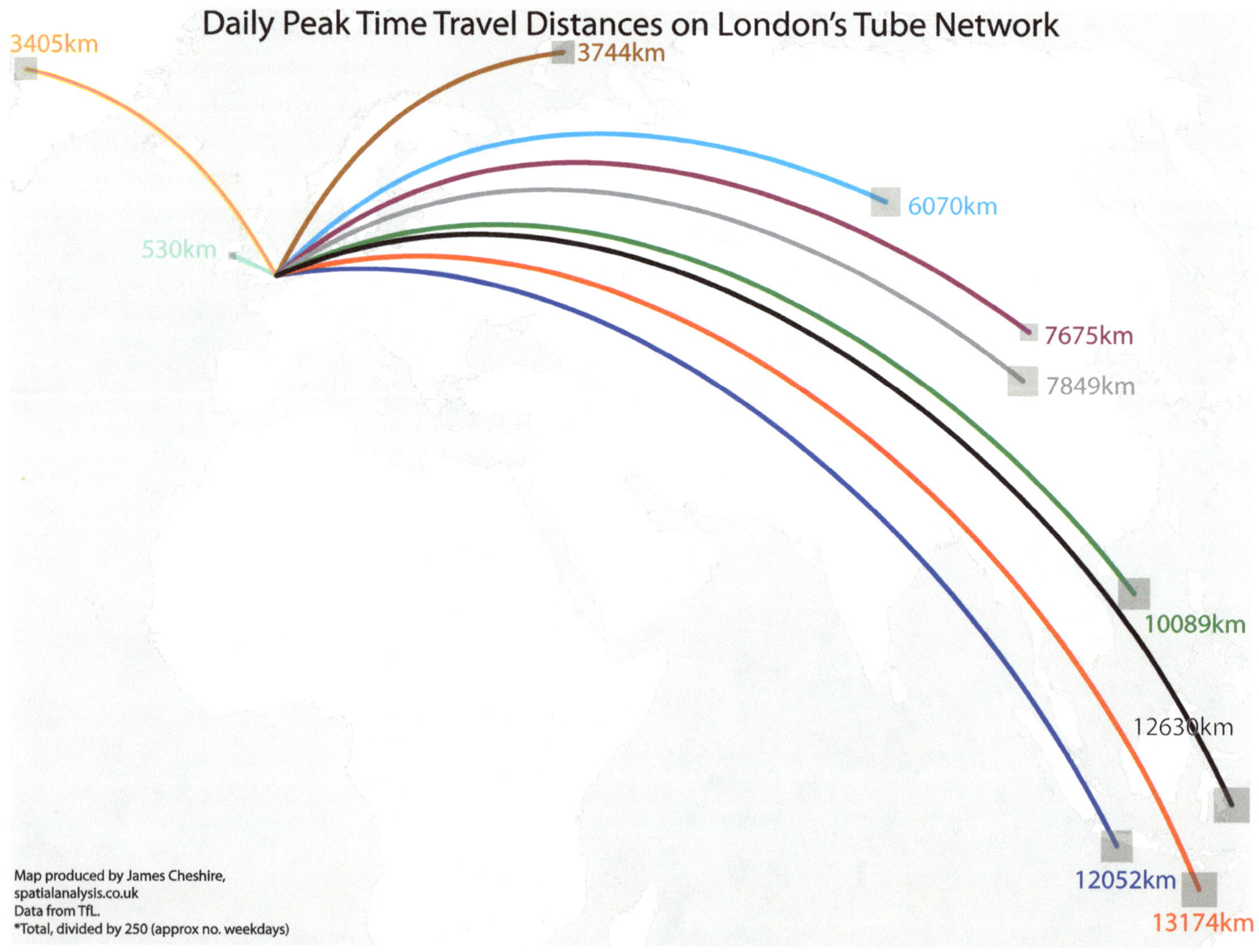

The London Tube Network: Daily Travel Distances by James Cheshire, lecturer at the UCL Centre for Advanced Spatial Analysis.

STYLE Maps that are entirely focused on flow arcs are most often depicted on a global scale, though it is possible to present connections at a medium or large scale if the data is spatially constrained yet rich enough to be meaningful. The connections between the points are typically represented by great circle arcs on a flat map or a globe. Great circle arcs represent the shortest distance between two points on the surface of a sphere. This is an interesting concept because it may not be immediately apparent to the cartographer that great circle arcs, especially on a flat-map projection, are appropriate.

Recent flow arc maps tend to have very dark blue, gray, or full-black backgrounds with brightly colored connection lines.[1] Light backgrounds with dark connection lines are also seen—especially if the flow arcs make up just one element in a more complex map composition. Because flow arc maps tend to be quite general, showing basic trends (even if a large amount of data is behind those trends), the background contextual information is kept simple. Political boundaries are not always present, for example. Indeed, when a large number of flow lines are mapped, sometimes they themselves form the background landmass structure.

Some flow arcs form connections, or hubs, as a natural by-product of their display. The connections are implied where the lines interconnect or merge, symbolized as small dots or as graduated circles where the size of the circle is related to the magnitude of connections at that point. These connections, at least the major ones, are often labeled as well, either with the location name or an explanatory note.

TECHNIQUE These maps are often created completely within R, a statistics software. Over-cluttering the map can be an issue when using massive datasets. In such cases, it is best to display only those connections that are of the highest importance. Line thickness and color can be used instead to show the magnitude and type of a variable. For example, a map of truck routes shows lines of varying widths depending on the number of tons associated with the flows. This produces an interesting effect where the places with the most flow have enough lines to become virtual *areas*, while the areas with sparser flow are depicted with lines emanating from the larger flow areas.[2] Animations of these maps could also be informative, using time series to show ebbs and flows of the variable as time passes or in periodic cycles.[3]

[1] A widely circulated example is Paul Butler's "Visualizing Friendships" map created for Facebook.

[2] See the U.S. Department of Transportation's Texas Truck Flows map, 1998.

[3] See *The Media Lab* by Stuart Brand, for perhaps the first written introduction to this idea. Brand, Stuart. *The Media Lab: Inventing The Future at M.I.T.* New York: Penguin, 1988.

Pseudo Natural

USAGE Maps with pseudo natural characteristics represent a departure from the more common flat-map backgrounds used in most web services and many printed products. As such, they provide a different, and often richer, user experience. For example, a typical flat-map background shows a national forest with a single, green hue. By contrast, a pseudo natural map background displays a natural forest with varying realistic shades of green and depending on the scale, individual trees of varying types accompanied by drop shadows. Using realistic shadowing and other naturalistic graphics provides a more realistic "world" in which to showcase map data.

An emerging concept that has not been fully implemented at the time of this writing is the transformation of flat maps from web services into pseudo natural maps on the fly. This type of pseudo natural map achieves its realism not through the typical cartographic workflow—manual placement of elements—but through automated algorithms. In this way, entire map services are created that can be used as backgrounds in webmaps of any scale and any region where there are corresponding flat-map services.

STYLE Pseudo natural maps employ real-life color schemes, realistic shadowing, and 3D effects.[1] A specific 3D technique called *bump mapping* is often used in pseudo natural maps, commonly for rendering realistic tree canopies, other vegetation, and beach sand. When a large amount of surface area needs to be textured with intricate shading, bump mapping is preferred over other techniques because it is faster. The algorithms that produce bump-mapped features do not alter the original surface; instead they add a surface of modeled shadows to the top of the feature. Texture, if not present on all mapped features, is at least a primary visual effect.

When natural features gradually give way to other natural features in real life, the pseudo natural map simulates it by gradually decreasing the density of one of the natural features while increasing the density of the adjacent one. This is distinctly different from normal flat maps that have sharply delineated feature boundaries. Because of this and their other

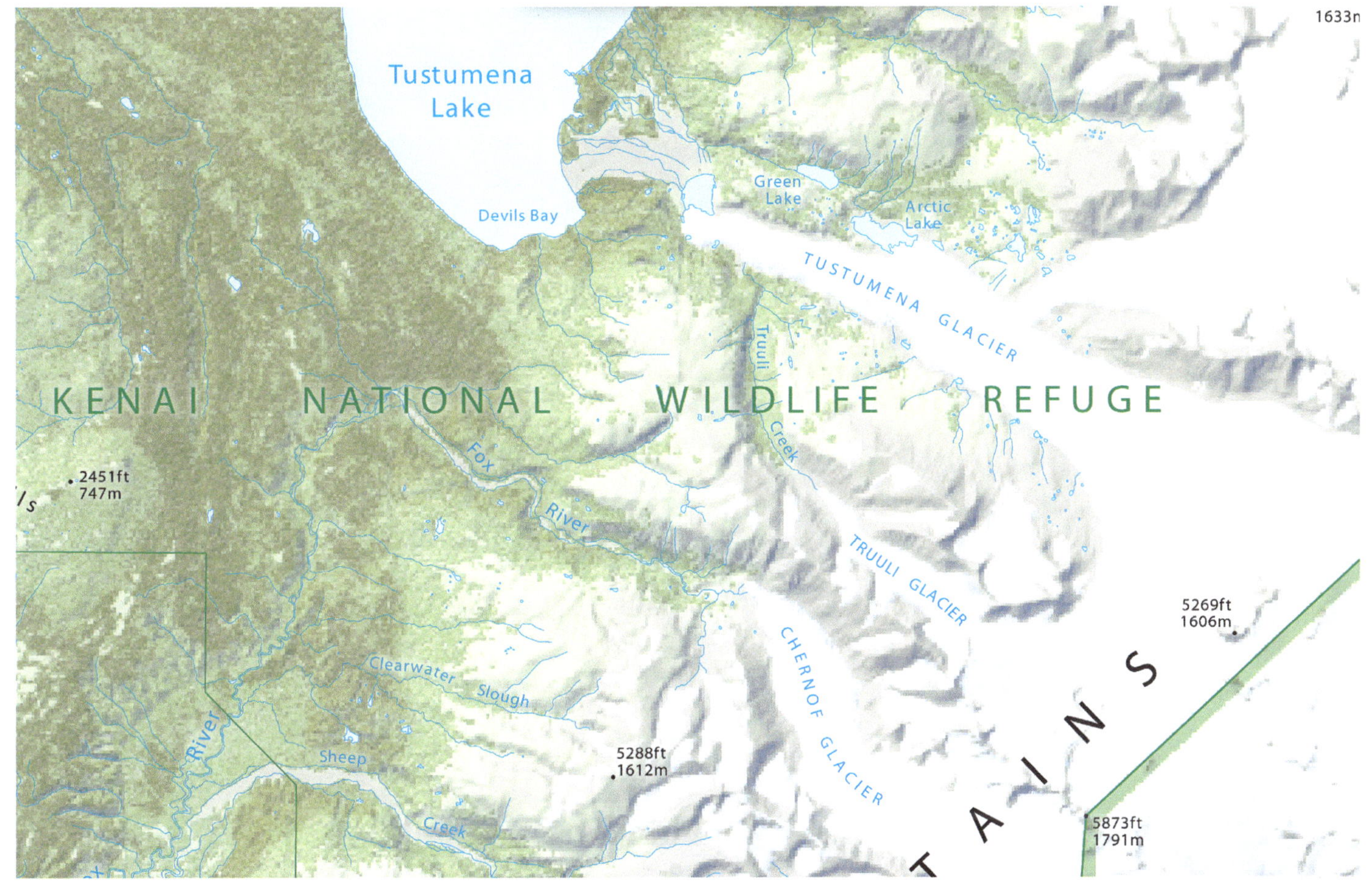

Kenai Fjords National Park, U.S. National Park Service.

realism devices, pseudo natural maps often give a faux aerial-photo feel to the display.

TECHNIQUE Texture contrast between features is useful on larger scale pseudo natural maps. For example, man-made features, such as roads, could be rendered flat and textureless relative to adjacent forest, beach, or meadows. The resulting contrast provides visual relief from what would otherwise be a completely textured display. To derive a naturalistic palette, capture color formulas from photographs of similar real-life landscapes to better estimate the exact amount of variance between the green hues of, for example, deciduous and coniferous trees.

[1]More information on this nascent technique is available from the Cartography and Geovisualization Group at Oregon State University.

Interactive

USAGE Interactive maps are digital maps either entirely created by the cartographer, or, more often, created as mash-ups on familiar basemaps such as Open Street Map or Google Maps.[1] Interactive maps provide a lot of spatial knowledge in a compact space compared to static or printed maps. From toggling layers, to pop-ups on mouseover, to increased detail on zoom, these maps enable a truly engaging experience. Used as teaching tools, decision-making tools, or navigational tools, these maps also span all conceivable subjects with a spatial component.

STYLE Interactive maps have the potential to provide complex data in a user-friendly interface, and indeed, this is how most interactive maps are designed. Conversely, they can be very simple—a two-toned store locator with a few pop-up pictures, for example, is also an interactive map.

How the user will interact with the map is just as important a style component as the cartographic aesthetic in these maps. The interactivity is often provided via semi-transparent navigational buttons (pan and zoom) placed on top of the map. Many interactive maps use basically the same menu styles because their users are already accustomed to them, thus making them the easiest to use. However, when it comes to layer switching, there are several common toggle methods in use, from tabs to radio buttons. These styles cannot all be discussed in the limited space available here and the cartographer is advised to learn the options via in-depth sources.

TECHNIQUE Requirement fulfillment and design must be addressed together, with neither outranking the other. In the implementation phase, it is not uncommon at the time of this writing to require the use of several-to-many technologies in order to create the needed functionality for a semi-complex to very-complex interactive map. This may change in the future, but until then, it is imperative that the interactive map designer be flexible in the implementation phase, modifying requirements and capabilities as needed.

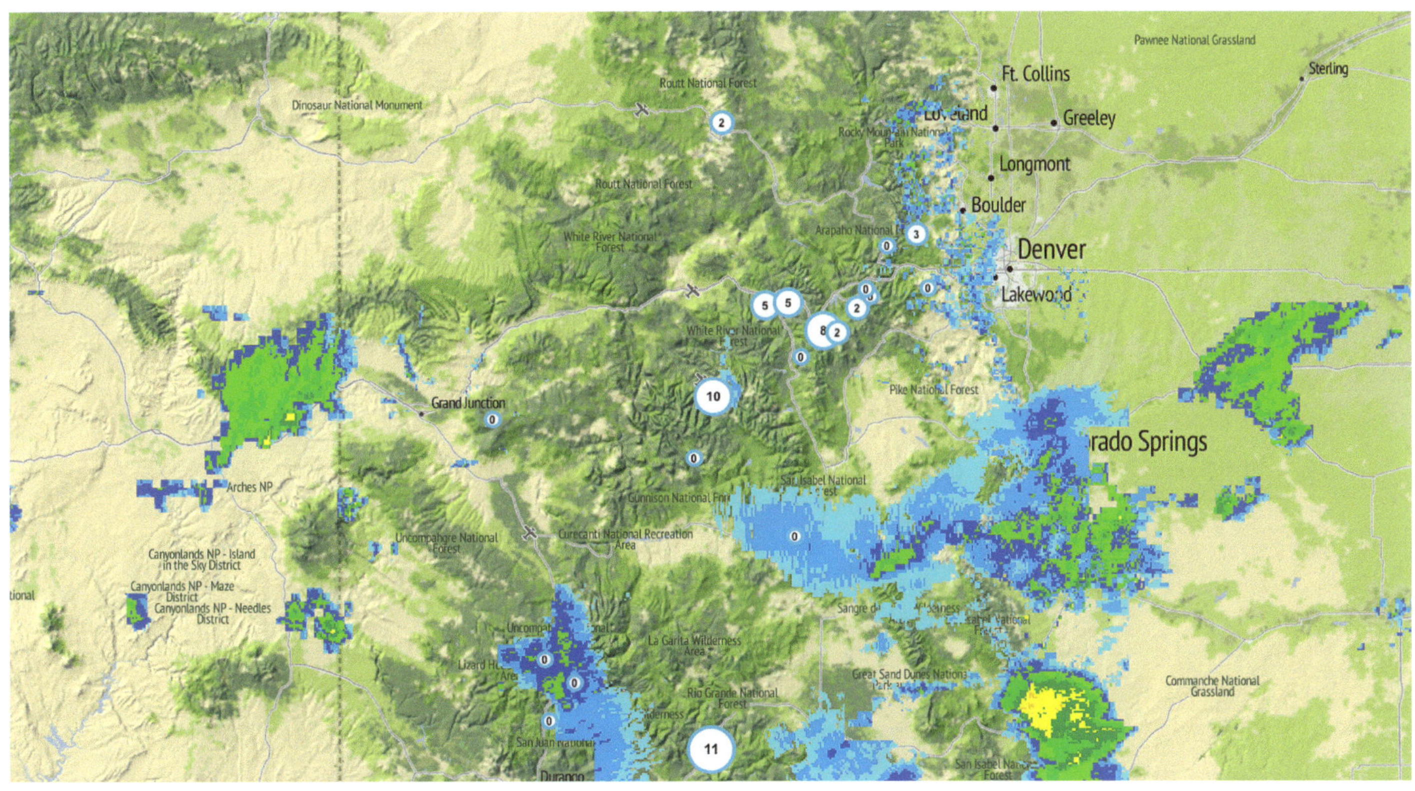

Freshy Map: Interactive Snow Conditions at Ski Resorts by Christopher Helm and Brendan Heberton.

With specific regard to the cartographic components of interactive maps, harmonious symbology for every possible layer combination can be a challenge. In extremely complex maps, the user's choices may need to be limited so as to disable the viewing of too many layers at the same time. The standard basemap is complex by itself. If there are many layers to be shown and the basemap information is not critical, a simpler basemap may be used instead.[2]

See also: Small Multiples, Animation

[1]Google Maps API and Open Street Map are common, but some new basemap services, including a unique watercolor basemap, have been added to the options more recently; see Stamen Design's examples online.

[2]See the Esri Light Gray Canvas map or Stamen Design's Toner for two options.

Animation

USAGE Animated maps enable the visualization of spatial data trends over space and time. In comparison to the small multiples pattern, animation is better for conveying overall trends, while static small multiples maps are better for alternating A multivariable animation example might be, for instance, a map of territorial acquisitions interspersed with battle locations and human migratory routes over time. Another interesting way to do a multivariable animation is to morph one variable into another.

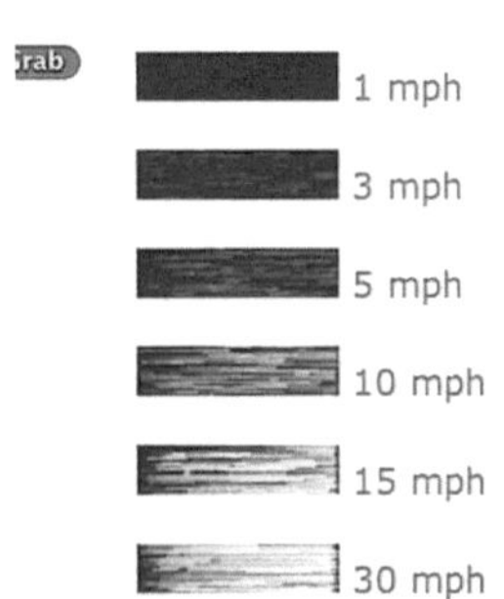

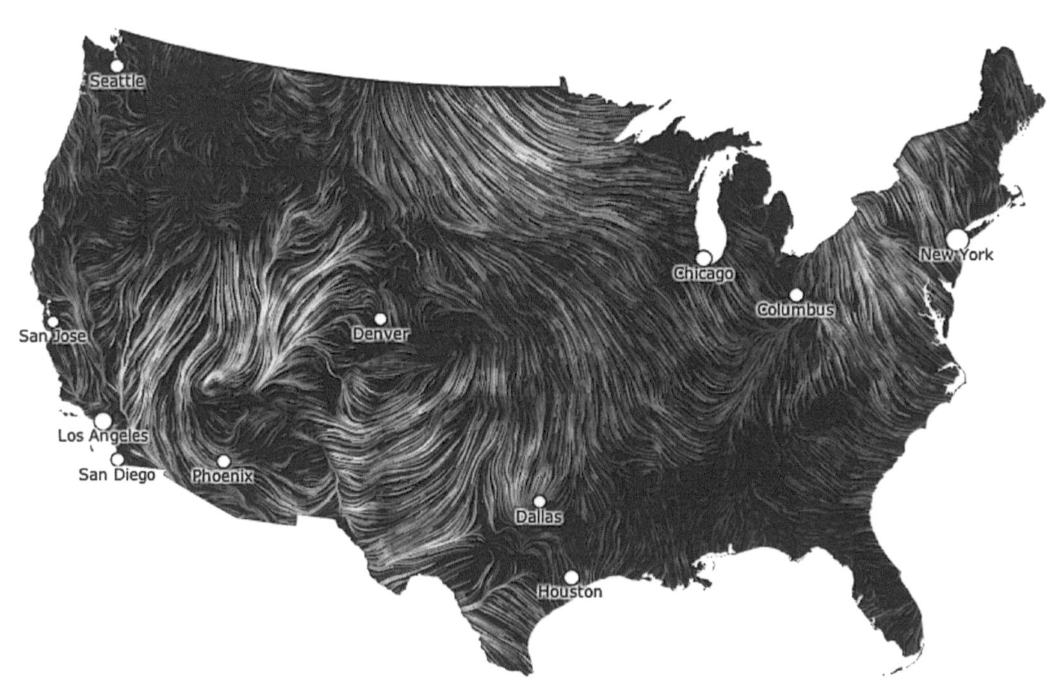

Near Real-time Wind Map by Fernanda Viégas and Martin Wattenberg, HINT.FM.

between individual points in time or scenes. Animations are useful for displaying a single variable changing over time, multiple variables changing over time, and spatial fly-throughs. Non-temporal animations include 3D fly-throughs or animated pan and zooms.[1] There are also movie-style animations of maps accompanied by audio, which are especially effective at teaching historical events.

STYLE Because animated maps can be very diverse in subject matter and contain any number of variables, a common style is hard to come by. Some animated maps allow user input via a time slider or a play-pause-stop button set, and some run continuously with no input. Likewise, very complex animated maps may allow the same capabilities as many webmaps, such as zoom and pan, where the animation still occurs but in a different scale or region. Often, any data that is changing over time is highly detailed, while the background information is left as simple as possible. Sometimes the exact spot that the animator wants the viewer to see is clearly demarcated via a large arrow pointing to the area that is changing or of particular interest.

TECHNIQUE Pace is an important consideration in animated maps. Different datasets will require different pacing, largely depending on whether the data trends are strong or weak. Strong data trends can be understood at a fast pace, while weak data trends require a slower pace. Because of this, an animation of multiple variables will be most effective if the user can control the pace so that the trends in each variable are discernible.[2] The length of time (duration) that each scene is displayed is also important. The shorter the duration of each scene, the more smoothly the animation will appear. Additionally, the date and time component can, and in many cases should, be shown on the map display, updating as the animation moves forward. If a slider is being used, the pointer can serve this function.

Another design factor is data density. Less dense data, such as just a few points that are far apart, might be symbolized as flashing points. More dense data, such as the wind isolines in this section's example, can be displayed as simple, thin lines. An ideal outcome for dense data is to provide both overall structure (Gestalt) and local details. As in the example shown here, the overall structure is elegantly displayed on the overview map of the United States, while the local details are available when the user zooms in.

See also: Interactive, Small Multiples

[1]See ICRAR's "6df Galaxy Survey fly through" on vimeo.com for an example of a 3D fly-through of the universe.

[2]See "A Comparison of Animated Maps with Static Small-Multiple Maps for Visually Identifying Space-Time Clusters," Amy L. Griffin, Alan M. MacEachren, Frank Hardisty, Erik Steiner, and Bonan Li, Annals of the Association of American Geographers, 96(4), 2006, pp. 740–753.

Complexity

USAGE Complexity in maps is anathema to many high-profile design writers who espouse simplicity instead. Complexity, though, has its place in many of the best-designed maps. Many of the examples in this chapter are effective complex maps. The quality of a map that is rich in information while still being readable and aesthetically up-to-par is apparent. For example, while it may seem like a bad idea to place a detailed parcel layer underneath a stream map, a parcel layer that has enough contrast with the stream layer (either much darker or much lighter, for example), can add a great deal of location information—as well as provide a richer-looking background for what would otherwise be a sparse map.

The reason that many design writers advise against complexity is that it requires a great deal of effort, and even then, the results are not a sure thing. Another reason is that complexity in design can include details that make the design more difficult to understand. For example, while elevation contours can be desirable in some instances, if there are too many contours for the scale of the map, and if they are too wide or too dark, they can easily contribute to unwelcome visual clutter. The lesson here is to ensure that inconsequential clutter is absent from the composition while simultaneously striving to provide as much necessary detail as possible.

The most successful maps achieve a Zen-like balance between being too complex and too simple. A too-complex map is confusing and off-putting. A too-simple map is uninformative and boring. As the design strategist Donald Norman puts it, "People prefer an intermediate level of complexity."[1]

STYLE Complexity is achieved by layering many elements onto the composition, while maintaining color and typographic harmony, appropriate contrast, and integration of insets. The style can also include maps with few data layers, where those layers contain multitudinous features—a map of buildings in a large city, for example. This latter type is generally easier to work with aesthetically.

TECHNIQUE A full cartographic workflow is needed with a complex map. The workflow starts with requirements gathered from all stakeholders, followed by trial sketches, and continues through a series of drafts and revisions based on formal design critique. Just as a book author needs copyeditors, proofreaders, and peer reviewers, so, too, does the maker of a complex map need copyeditors, proofreaders, and peer reviewers.

In complex maps, it can be difficult to achieve a visual focus, if one is desired. One technique is to include the complexity only in those areas that need focus, while maintaining a simpler look in the non-focus areas. For example, most map layers can be clipped to a central feature such as a watershed, while only a few—such as topography, perhaps—are left unclipped and therefore run across the whole page (see also Focal Point). Another technique is to intentionally decrease the complexity around the focal point for a certain distance. This provides a natural separation between the focal point and the rest of the map in the form of white space. For example, labels, points, and lines can be decreased or eliminated immediately surrounding a large city on a regional map. If the cartographer is against eliminating features or labels, they can at least become visually less apparent by decreasing label point size, desaturating color, and narrowing line widths.

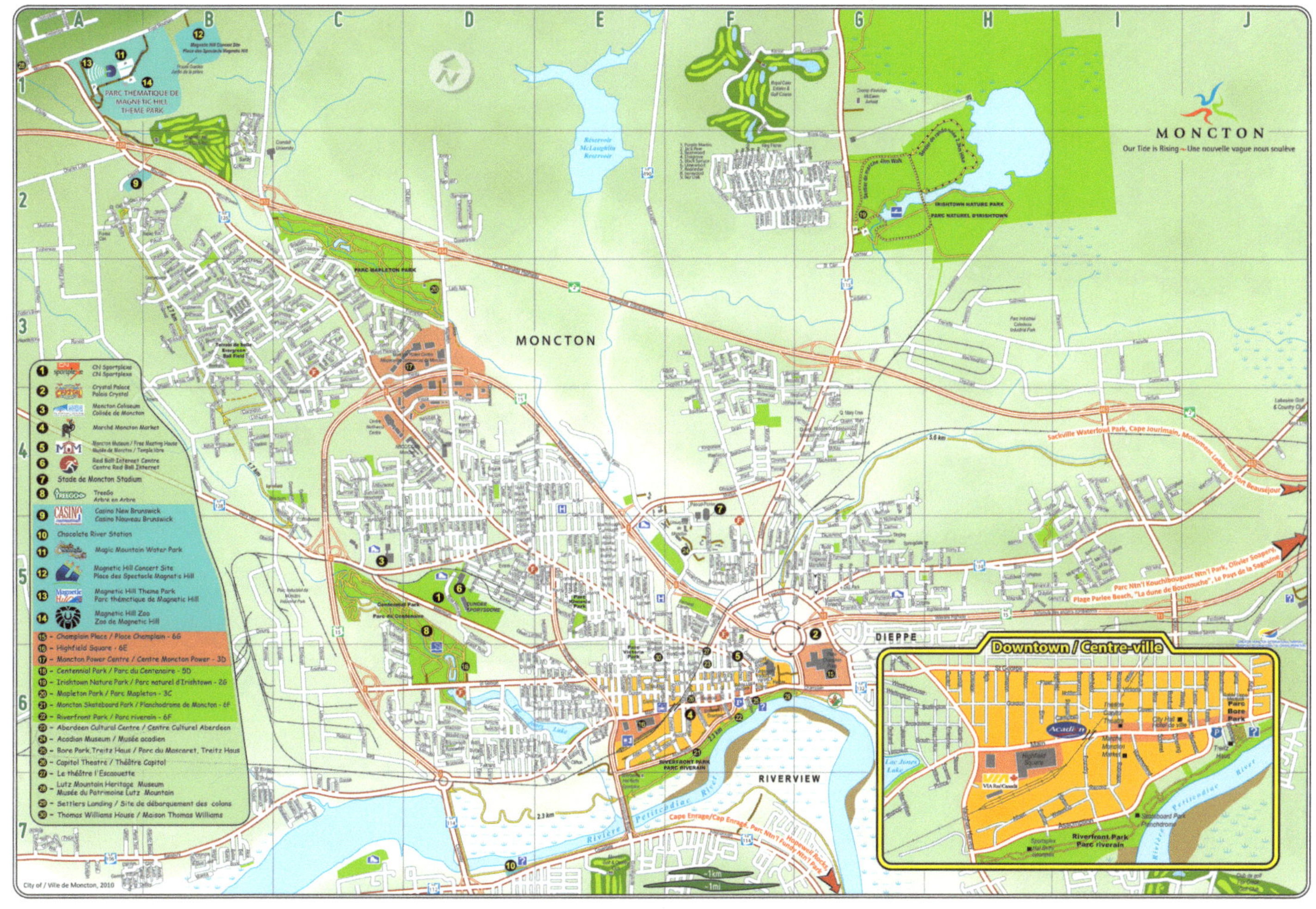

Moncton Tourist Map by Daniel Gray, City of Moncton.

To stay away from overly complex design, one technique is to minimize variation in one of the map elements. For example, a map with parcel outlines, place-name labels, canal and river lines, roads, and a heat map overlay can achieve a good middle-ground between simple and complex by minimizing the number of colors used. The parcel outlines might be gray, the heat map might be just a few shades of a single color, and the other features might all be rendered in a similar, neutral hue. Even if the colors are rich, if there are just few of them, the map will be more likely to appear cohesive and comprehensible.

[1]Norman, Donald A. *Living with Complexity.* Cambridge, MA: The MIT Press, 2011.

Afterword

The current trend in map design is toward a richness of data presentation rarely seen in the more business-minded maps of the past decades. There is now a greater variety in map forms: from serious to humorous, from simple to detailed, and from muted to vibrant. Part of the reason for this is that there is now a greater variety of map makers and part is because of greater access to better software.

Cartographers are not the only ones making good maps these days. Analysts, statisticians, and geographers who have taken the time to learn the principles of map design in addition to their regular work are also creating enticing, communicative maps. Additionally, an influx of non-data oriented map makers are dabbling in map design, many with vastly different ideas than traditional map makers. These professionals are approaching their products from a fresh angle, creating interesting and novel maps.

The current richness and variety in maps is also a function of the greater availability of advanced mapping and graphic design software, which generally has more features, is easier to use, and is easier to obtain than in the past. Free and open source software, for example, is now a viable option for some mapping needs, especially in the web domain. Commercial software's ease of use has increased and cartographic output of GIS software, in particular, is gaining in quality. Between the higher professional competition and better software, everyone making maps must now be mindful of staying current with the latest cartographic techniques.

Cartography is just as important now as it has ever been. With the emergence of Big Data in the coming decades, a combination of mathematical savvy and design savvy will further benefit the cartographer. Well-designed maps enable data professionals, in particular, to glean interesting and world-improving conclusions from the massive datasets being collected, and to effectively communicate those conclusions to the populace.

This book has explored advanced cartography and provided tools and information for producing the high-level maps required in today's innovative environment: color palettes, typeface comparisons, and existing, proven patterns of map composition. The information presented in this book, along with the more fundamental cartography theory put forth in the author's first book, *GIS Cartography: A Guide to Effective Map Design*, will arm cartographers with the tools they need to perform at the top of their field, producing maps that are informative, inspired, and original.

Printed in the USA
CPSIA information can be obtained
at www.ICGtesting.com
LVHW060238141223
766412LV00003B/59

* 9 7 8 0 6 1 5 4 6 7 9 4 8 *